T0233552

Global Crisis and Sustainability Technologies

Global Crisis and Sustainability Technologies

Kenji Uchino
Pennsylvania State University, USA

 **World Scientific**

NEW JERSEY · LONDON · SINGAPORE · BEIJING · SHANGHAI · HONG KONG · TAIPEI · CHENNAI · TOKYO

Published by

World Scientific Publishing Co. Pte. Ltd.

5 Toh Tuck Link, Singapore 596224

USA office: 27 Warren Street, Suite 401-402, Hackensack, NJ 07601

UK office: 57 Shelton Street, Covent Garden, London WC2H 9HE

Library of Congress Cataloging-in-Publication Data
Names: Uchino, Kenji, 1950–   author.
Title: Global crisis and sustainability technologies / Kenji Uchino
    (Pennsylvania State University, USA).
Description: New Jersey : World Scientific, 2016. | Includes bibliographical references and index.
Identifiers: LCCN 2016014276 | ISBN 9789813142299 (hc : alk. paper) |
    ISBN 9789813143722 (pbk : alk. paper)
Subjects: LCSH: Technology--Social aspects. | Environmental protection. |
    Green technology. | Sustainable engineering. | Industrial safety.
Classification: LCC T14.5 .U155 2016 | DDC 338.9/26--dc23
LC record available at http://lccn.loc.gov/2016014276

British Library Cataloguing-in-Publication Data
A catalogue record for this book is available from the British Library.

Desk Editors: Herbert Moses/Amanda Yun

Typeset by Stallion Press
Email: enquiries@stallionpress.com

Printed in Singapore

To my wife, Michiko,
who constantly encourages me

Preface

I have been a so-called "Navy Ambassador to Japan" (my official title was Associate Director at the Asia Office of the US Office of Naval Research) for four years until July 2014. When I accepted this special assignment from the government, I initially expected to be assigned relatively relaxed job tasks because I had not taken any official sabbatical leave for more than 20 years. However, less than half a year after my arrival in Tokyo, I found myself having to face the Big Earthquake and consequent Fukushima Daiichi Nuclear Power Plant meltdown on March 11, 2011, which changed my job-tasks dramatically; I became very busy, assisting in the US rescue operation known as Operation "*Tomodachi* (friendship)", on a tight schedule. In parallel, I had an opportunity to get involved in setting up multiple international R&D related agreements between the US Department of Defense and Japanese governmental institutes, including those pertaining to the rescue technology projects relating to these catastrophic disasters.

It was while carrying out these diplomatic tasks that I confirmed the urgent necessity for politically-initiated technology development in Japan. Historically, it was the Japanese government who set the four-Chinese-character slogans that encouraged researchers to pursue research along a particular direction. "重, 厚, 長, 大 (heavier, thicker, longer, and larger)" was the first slogan. Launched in the 1960s, it was aimed at infrastructure recovery post-World War II. At that time, I inevitably tried to become a "huge dam" engineer owing to my choice to join the Electrical Engineering department at a Japanese university. Then, in the 1980s, a completely opposite slogan, "軽, 薄, 短, 小 (lighter, thinner, shorter, and smaller)," was started, with the aim to strengthen the country's economic power. It was then that Japanese semi-conductor industries begun to chase leading US companies aggressively. I started my work on compact piezoelectric actuators for "micromechatronic" (a term I created) applications in the late 1970s following this trend, as an assistant professor in a Japanese university. When the long economic depression started in the late 1980s in Japan, I was recruited by Penn State University, USA, and thus immigrated to the US. Since the time we entered the 21st century, with the maturity of economic power in Japan, the requirements and expectations for science and technology and engineering seem to have changed again.

I would now like to propose a new four-Chinese-character slogan for the 21st century: "cooperation, protection, reduction, and continuation (協, 守, 減, 維)," particularly to Japan. International *cooperation* and global collaboration in the standardization of internet systems became essential for accelerating mutual communication. The US–Japan Agreement in the development of "Rescue Robots" for crises was one of the urgent tasks in which I was involved. The Kyoto Protocol in December 1997 was a trigger to more wide international agreements linked to the United Nations Framework Convention on Climate Change in order to reduce greenhouse gas emissions. It is a symbolic global regime for determining the direction of research in the 21st century. Technology development is essentially constrained by these global regimes. The *protection* of sovereign territory and the environment from enemies, natural disasters, and the spread of infectious diseases is mandatory. In addition to terrorist attacks, HID, Bird Flu, and in particular, EBOLA, are now also examples of such worldwide challenges. Cyber protection and counter-weapons also need to be developed urgently — more so today in the wake of some recent high-profile hacking attacks, such as the hacking incident involving SONY Pictures. *The reduction* of the use of toxic materials (such as lead, heavy metals, dioxin) and resources, as well as a reduction in energy consumption are also key to the continuity of society. It is important to promote the idea of creating a sustainable society. Even in my research area, the long-term material champion, lead zirconate titanate (PZT), may be regulated in several years by the Restriction of Hazardous Substances Directive (RoHS) due to its "Pb (lead)" inclusion. We recognize various toxic compounds such as indium gallium arsenide even in semiconductor materials. Thus, current material researchers need to seek alternative (environmentally-friendly) materials to replace toxic ones. In parallel, we should consider how to eliminate/neutralize already existing toxic and/or hazardous materials such as asbestos, dioxins, polychlorinated biphenyl (PCB) and some biological hormones. I am proud that my invention, the "multilayer piezoelectric actuator" has become one of the key technologies contributing to this by reducing NO_x or SO_x in recent diesel engine automobiles. The energy issue is also one of the main themes to ensure in future sustainability. Because of some anxieties regarding the indefinite maintenance of nuclear power plants, particularly after the Fukushima Daiichi Power Plant accident, renewable energy researches are booming now. My company's present renewable energy products include piezoelectric energy harvesting devices that harvest energy from mechanical noise vibrations.

Due to the key issues given in the examples above, I decided to write an overview about "*politico-engineering*" (politically-initiated engineering) for current times — specifically, of technologies for handling *crises* (such as natural disasters, infectious diseases, enormous accidents, terrorist/criminal incidents, war/territorial invasion) and for *societal sustainability* (through the elimination of toxic materials, and the employment of renewable energy), as well as risk management in international relationships. Because of my experience, the coverage of this book is limited mostly to the US and Japan, and a little on China. These three are the current leading countries (in gross domestic product; GDP) in the world with similar country power growth trends (compared to Japan, the US is 20 years ahead and China 30 years behind). Also, many of the topics in S&T technical development cited in this book come from my area of expertise, "piezo-electric actuators." However, I believe that all case studies can be extended to and be instructive to a wider group of readers.

This book covers:

Chapter 1: Background of Politico-Engineering, where I will discuss the four forces (social/cultural/religious, technology/science, economics, and politics/law) influencing progress in engineering, from a historical viewpoint.

Chapter 2: Japanese and Global Technology Trend Change, where three phases, Initial Growth, Rapid Growth, and Growth Maturing, are described with politically-initiated development slogans.

Chapter 3: Global Politics in Engineering, where general global regime making is discussed with the simplest game theory, and the key words, Cooperation, Protection, Reduction, Continuation, are introduced as case studies.

Chapter 4: Categorization of Politico-Engineering, where the differences between normal and crisis technologies are compared in order to identify which are products of politically-initiated engineering.

Chapter 5: Crisis Technology, where we take a look at case studies of natural disasters, infectious/contagious diseases, enormous accidents, intentional (terrorist/criminal) incidents, and external and civil wars/territorial invasions.

Chapter 6: Sustainability Technology, where we will cover power and energy, rare materials, foods (as biofuel), toxic materials, environmental pollution, and energy efficiency.

Chapter 7: Risk Management, where the basics of risk management are described, followed by individual, non-national and national risk/crisis management, including an "Anti-Terrorism Training Program" as a case study.

Chapter 8: Advanced Game Theory, where game theory is used to understand international risk/crisis management, global regime policies and strategic decision-making in international relationships.

Chapter 9: Concluding Remarks.

I have been a university professor for 41 years, a company executive (president/vice president) for 22 years and a government officer both in Japan and US for 7 years so far. I used to purely be an engineer specializing in "micro-mechatronics" (piezoelectric actuators in particular) until my mid-40s. After immigrating to the US, I started my own companies, to commercialize the technologies I had invented in the universities. At the same time, I concurrently took on an MBA degree and started teaching "entrepreneurship for engineers" in various universities using the strategies behind the development of piezoelectric devices as case studies. Thanks to a kind assignment by the US Office of Naval Research, I had a chance to expand my experience by entering into diplomatic/political areas. I greatly appreciate my former colleagues from the ONR Global — Asia Office (Tokyo), for supporting me in this field. Without this opportunity, this book would not have crystallized. A special appreciation is extended to Ms. Amanda Yun, Senior Editor at World Scientific Publishing, for her thorough linguistic improvement of my manuscript.

This book's content is based on the lecture notes I wrote on "Crisis Technology" for the class I taught in Keio University, Japan, during my Tokyo stay, and on the lecture notes for my "Politico-Engineering" course held in the Penn State University, USA, after my return in 2014. As far as I have searched, I have found very few books focusing on this interdisciplinary area between engineering and global politics/international relationship, thus motivating me to write this book. The coverage of this book is naturally, not exhaustive, nor should it be regarded as such as it is impossible to cover the topics comprehensively. Nonetheless, this book may be especially suitable for undergraduate and graduate students studying engineering and law, as well as for active researchers who are interested in becoming Science & Technology (S&T) program officers, or who have an interest in intellectual properties, S&T policy making, global regimes, etc. Critical comments from readers on this book are most welcome.

April 1, 2016

Kenji Uchino
Professor, Electrical Engineering Director,
International Center for Actuators and Transducers State College,
Pennsylvania State University, USA

Author's Biodata

Kenji Uchino, one of the pioneers in piezoelectric actuators, is *Founding Director of* the *International Center for Actuators and Transducers* and Professor of Electrical Engineering and Materials Science and Engineering at Pennsylvania State University, USA. He was *Associate Director* (*Global Technology Awareness*) *at The US Office of Naval Research —* Global (Tokyo Office) through the IPA (Intergovernmental Personnel Act) from 2010 to 2014. He was also the *Founder and Senior Vice President and CTO of Micromechatronics Inc.*, at State College, Pennsylvania, USA. After being awarded a PhD degree from Tokyo Institute of Technology, Japan, he became Research Associate/Assistant Professor (1976) in the Physical Electronics Department at the same university. He then joined Sophia University, Japan, as Associate Professor in the Physics Department in 1985. He was later recruited by Pennsylvania State University, USA, in 1991. He was also involved in the Space Shuttle Utilizing Committee in NASDA, Japan, from 1986–1988, and was Vice President of NF Electronic Instruments, USA, from 1992–1994. He was the Founding Chair of the Smart Actuators/Sensors Committee, Japan Technology Transfer Association, sponsored by Japan's Ministry of Economics, Trading and Industries, from 1987–2014, and has been a long-term session chair of the International Conference on New Actuators, Messe Bremen, Germany since 1997. He was also Associate Editor for the *Journal of Advanced Performance Materials*, the *Journal of Intelligent Materials Systems and Structures*, and the *Japanese Journal of Applied Physics*. Uchino has also served as an elected member of the *Administrative Committee of IEEE* Ultrasonics, Ferroelectrics and Frequency Control (1998–2000) and as *Secretary of the American Ceramic Society*, Electronics Division (2002–2003).

His research interests are solid state physics — especially in ferroelectrics and piezoelectrics, including basic research on theory, materials, device designing and fabrication processes — as well as application developments of solid state actuators/sensors for precision positioners, micro-robotics, ultrasonic motors, smart structures, piezoelectric transformers and energy harvesting. K. Uchino is known as the discoverer/inventor of the following famous topics: (1) lead magnesium niobate (PMN)-based electrostrictive materials, (2) cofired multilayer piezoelectric actuators (MLA), (3) superior piezoelectricity in relaxor-lead titanate-based piezoelectric single crystals (PZN-PT), (4) photostrictive phenomenon, (5) shape memory ceramics, (6) magnetoelectric composite sensors, (7) transient response control scheme of piezoelectric actuators (Pulse-Drive technique), (8) micro ultrasonic motors, (9) multilayer disk piezoelectric transformers, and (10) piezoelectric loss characterization methodology. His ongoing research projects are also in the areas listed above, especially in topics (8), (9), and (10) most recently. He has authored 530 *papers*, 68 *books and 31 patents* in area of the ceramic actuators so far. Forty-one papers/books among his publications have been cited more than 100 times, leading to his average *h-index* score of 66. The total number of times his work has been cited stands at ~23,000, and his annual average citation number, 470, is also considered very high in the College of Engineering.

He also holds an MBA degree, awarded by St. Francis University, USA (2008), and has authored a textbook, *Entrepreneurship for Engineers*, for the Smeal College of Business, Penn State University, USA. He is a *Fellow of the American Ceramic Society* since 1997, a *Fellow of IEEE* since 2012, and is also a recipient of 28 awards, including the *International Ceramic Award from the Global Academy of Ceramics* (2016), *IEEE-UFFC Ferroelectrics Recognition Award* (2013), *Inventor Award from the Center for Energy Harvesting Materials and Systems, Virginia Tech* (2011), *Premier Research Award from The Penn State Engineering Alumni Society* (2011), the Japanese Society of Applied Electromagnetics and Mechanics Award on Outstanding Academic Book (2008), SPIE (Society of Photo-Optical Instrumentation Engineers), *Smart Product Implementation Award* (2007), *R&D 100 Award* (2007), ASME (American Society of Mechanical Engineers) *Adaptive Str uctures Priz e* (2005), *Outstanding Researc h Award from the Penn State Engineering Society* (1996), Academic Scholarship from Nissan Motors Scientific Foundation (1990), Best Movie Memorial Award at the Japan Scientific Movie Festival (1989), and the Best Paper Award from the Japanese Society of Oil/Air Pressure Control (1987). He is also one of the founding members of the *Worldwide University Network*, which encourages linkage between UK and US universities since 2001.

Contents

Preface   vii
Author's Biodata   xi

Chapter 1   Background of Politico-Engineering   1

    1.1 Prologue   1
        1.1.1 Game Theory   1
        1.1.2 Decision-Making   2
        1.1.3 Influence of Advanced Technology   4
    1.2 Four Influencing Factors on Engineering   5
    1.3 Socio-Engineering   6
        1.3.1 Alchemy   6
        1.3.2 Copernican   8
        1.3.3 Battle between Science and Religion   10
    1.4 Techno-Engineering   14
        1.4.1 Newton and Descartes   15
        1.4.2 Boyle   16
        1.4.3 New Mathematics   17
        1.4.4 Franklin   18
        1.4.5 Coulomb   20
        1.4.6 Huygens   21
        1.4.7 Discovery of Piezoelectricity   21
    1.5 Econo-Engineering   22
        1.5.1 Industrial Revolution   22
        1.5.2 Spinning and Weaving Machines   23
        1.5.3 Steam Engines/Turbines   24
        1.5.4 Machine Tools   25
        1.5.5 American Mass-Production System   27
    1.6 Politico-Engineering   28
        1.6.1 École Polytechnique   29
        1.6.2 The US Naval Research Laboratory   30

1.6.3 Technology Developments during World War II   32
1.6.4 Piezoelectric Application History   35
1.7 Summary   38
1.8 Problem   39
References   39

Chapter 2 Japanese and Global Technology Trend Change   41

2.1 Country Power Growth   41
    2.1.1 The USA   42
    2.1.2 Japan   43
    2.1.3 China   43
2.2 Initial Growth 1960s — Domestic Politics   43
    2.2.1 Heavier, Thicker, Longer, Larger   43
    2.2.2 Tokyo Summer Olympics   44
    2.2.3 Industrial Pollution   44
2.3 Rapid Growth 1980s — Econo-Politics Revival   46
    2.3.1 Lighter, Thinner, Shorter, Smaller   46
    2.3.2 Piezoelectric Devices   47
    2.3.3 Cost Minimization Strategy   48
    2.3.4 Beautiful, Amusing, Tasteful, Creative   52
    2.3.5 Sustainability Problems   52
2.4 Growth Maturing 2000s — International Politics   53
2.5 Summary   53
References   54

Chapter 3 Global Politics in Engineering   57

3.1 Levels of Systematization   57
3.2 Global Regime — Fundamentals   59
3.3 Global Regime — Categorizations   60
    3.3.1 Payoff Table   60
    3.3.2 Non-Regime Type Games   63
    3.3.3 Regime Type Games   65
3.4 Global Regime — Case Studies (Cooperation,
     Protection, Reduction, Continuation)   69
    3.4.1 Cooperation   69
    3.4.2 Protection   72
    3.4.3 Reduction   73
    3.4.4 Continuation   75
3.5 Summary   76
3.6 Problem   78
References   78

Chapter 4   Categorization of Politico-Engineering   79

   4.1 Crisis Technologies   79
        4.1.1 Natural Disasters   80
        4.1.2 Epidemics/Infectious Diseases   80
        4.1.3 Enormous Accidents   81
        4.1.4 Intentional Accidents   82
        4.1.5 War and Territorial Aggression   85
   4.2 Sustainability Technologies   88
        4.2.1 Power and Energy   88
        4.2.2 Rare Materials   89
        4.2.3 Food   91
        4.2.4 Toxic Materials   92
        4.2.5 Environmental Pollution   93
        4.2.6 Energy Efficiency   94
        4.2.7 Bio/Medical Engineering   95
   4.3 Paradigm Shift of Crisis Technology Development   96
        4.3.1 Automobile Diesel Injection Valve   97
        4.3.2 Deformable Mirror   98
   4.4 Summary   100
   References   101

Chapter 5   Crisis Technology   **103**

   5.1 Natural Disaster   103
        5.1.1 Earthquake/Volcanic Eruption   104
        5.1.2 Tsunami   110
        5.1.3 Tornado   111
        5.1.4 Hurricane/Typhoon/Flood   112
        5.1.5 Lightning   113
   5.2 Epidemic/Infectious Disease   114
        5.2.1 Detection of Infectious Disease Germs   115
        5.2.2 Disinfection of Disease Germs   115
        5.2.3 Vaccine/Medicine   116
   5.3 Enormous Accident   117
   5.4 Intentional Accident   119
        5.4.1 Acts of Terrorism, Criminal Activities   119
        5.4.2 Cyber Threats   122
   5.5 Civil War, War, and Territorial Aggression   123
        5.5.1 "Green" Weapons — Environment-Friendly Weapons   123
        5.5.2 Warfare Gadgets   125
        5.5.3 Unmanned Vehicles   130

5.6 Naval S&T Strategic Plan 132
  5.6.1 Office of Naval Research 132
  5.6.2 Naval Research Laboratory 133
  5.6.3 Naval Research Enterprise 133
  5.6.4 Naval S&T Strategic Plan 134
5.7 Summary 137
5.8 Problem 137
References 138

Chapter 6 Sustainability Technology 141

6.1 Power and Energy 141
  6.1.1 Fossil Fuel 142
  6.1.2 Zero-Carbon Emission 146
  6.1.3 Hydroelectricity 147
  6.1.4 Nuclear Power Plant 148
  6.1.5 Renewable Energy 152
  6.1.6 Fuel Cell/Batteries 157
  6.1.7 Electricity Grid 158
6.2 Rare Material 160
  6.2.1 Rare-Earth Metals 160
  6.2.2 Rare Metals 160
6.3 Food/Water 161
  6.3.1 Food 161
  6.3.2 Water 161
6.4 Toxic Material 163
  6.4.1 Restriction of Toxic Materials 163
  6.4.2 Replacement Materials 164
  6.4.3 Elimination of Hazardous Materials 167
6.5 Environmental Pollution 168
  6.5.1 Reduction of Contamination Gas 168
  6.5.2 Protection of River/Sea Pollution 169
6.6 Energy Efficiency 169
  6.6.1 Energy Density 169
  6.6.2 Loss Mechanisms in Smart Materials 170
6.7 Bio/Medical Engineering 171
  6.7.1 Artificial Fertilization System 172
  6.7.2 Medical MicroPump 172
  6.7.3 Micro-Monitoring and Surgery 173
  6.7.4 Drug Delivery 174

    6.8 Summary   175
    References   176

Chapter 7   Risk Management   **179**

    7.1 Level of Risk Management   179
        7.1.1 Threat Provider View   180
        7.1.2 Threat Receiver View   183
    7.2 Japanese R&D Ethics   185
        7.2.1 Military vs. Civilian Applications   185
        7.2.2 Medical Ethics   186
        7.2.3 Ecological Ethics   187
    7.3 Individual Survival Training   187
        7.3.1 Anti-Terrorism Training Program   189
        7.3.2 Kidnap Escaping Program   193
    7.4 Non-Nation Group: Crisis Management   199
        7.4.1 Civil War   199
        7.4.2 Company Risk Management   200
        7.4.3 Military FM 3-0 Operations Manual   205
    7.5 National Crisis Management   211
        7.5.1 International Terrorism   211
        7.5.2 Natural Disaster   214
        7.5.3 Nuclear Power Plant Accident   216
    7.6 Summary   232
    7.7 Problem   232
    References   233

Chapter 8   Advanced Game Theory   **235**

    8.1 Four Types of International Security Systems   235
        8.1.1 Competitive Security System (Type A)   235
        8.1.2 Controlled Power Balance (Type B)   236
        8.1.3 Crisis Response System (Type C)   238
        8.1.4 Inclusive Security System (Type D)   240
    8.2 Office of Naval Research Global   244
        8.2.1 Foundation of ONR Global   244
        8.2.2 Strategy of ONR Global   245
    8.3 Advanced Game Theory   248
        8.3.1 Merchant Rule in the Middle Ages   248
        8.3.2 Super Game Theory — Sequential Decision-Making   249
        8.3.3 Multiple-Player Cooperative Game   254

8.4 International Relation Case Studies   259
   8.4.1 Cuban Nuclear Missile Crisis   259
   8.4.2 Operation "Tomodachi"   267
   8.4.3 Cyber War   273
8.5 Summary   277
References   278

Chapter 9   Concluding Remarks   281

9.1 Urgent Requirements for Crisis/Sustainability
    Technologies   281
9.2 Hard Power vs. Soft Power   282
   9.2.1 The US Monopolar State   282
   9.2.2 Smart Power Strategy   282
9.3 Epilogue   284
References   286

Index   287

Chapter 1

Background of Politico-Engineering

1.1 Prologue

No era can be found when developments in engineering, science, and technology directly influenced a country's military and economic power more than the 20th century and after [see technology trends discussed in Chapters 1 and 2]. In the 1970s, the Organization of the Petroleum Exporting Countries (OPEC) was an internationally permitted "industrial cartel" or "collusion" — a system which would have been regarded as illegal within many nations. The Islamic Republic of Iran, Iraq, Kuwait, Saudi Arabia, and Venezuela, later joined by Qatar, Indonesia, Libya, the United Arab Emirates, Algeria, Nigeria, Ecuador, Gabon, and Angola, had set up an international regime with a mission "to coordinate and unify the petroleum policies of its Member Countries and ensure the stabilization of oil markets in order to secure an efficient, economic, and regular supply of petroleum to consumers, a steady income to producers and a fair return on capital for those investing in the petroleum industry" [see the section on global regimes in Chapter 3]. I experienced the first "Oil Shock" (1978–1980) in Pennsylvania first hand, by paying a high price that was almost triple of what I had previously paid at most gas stations, because of OPEC's intentional oil production reduction tactic. It was in the midst of experiencing the effects of this international cartel that I realized that *a global regime seemed to be set in ways that differed* from how normal domestic regulations/rules are set.

1.1.1 Game Theory

Though OPEC is an international cartel, it is still possible for some of its member countries to not keep to the oil-production reduction agreement in a hope to increase their distribution share, owing to the global economic depression unfolding at the time. Let us consider a hypothetical example case involving the interrelationship of Saudi Arabia ("Oligopoly" and the OPEC leader country) vs. Kuwait (one of the smallest contributors) in the 1970s. We can easily imagine

1

that if Kuwait agrees with the fixed higher price by controlling (i.e., reduction in) oil production in cooperation with Saudi Arabia, both will earn reasonable profits ($300M and $100M, respectively) because of the oil price increase (i.e., the so-called Oil Shock). Note that the profit of Saudi Arabia should be smaller than that of Kuwait, because oil-consumer countries may switch their supply source from Saudi Arabia (the OPEC leader) to outside suppliers in order to protest or as revenge on the intentional cartel's production control initiated by Saudi Arabia. However, if Kuwait reneges and continues producing pre-agreement amounts of oil, the profit of Kuwait will increase up to $500M with the increment in market share. Accordingly, Saudi Arabia's profits may become negative, –$1000M, if Saudi Arabia keeps to the production reduction strategy. Table 1.1 summarizes profit and deficit for Kuwait and Saudi Arabia according to the oil production strategy. This is called the *Payoff Table*, which includes two players/actors, (1) Kuwait and (2) Saudi Arabia, and two action items, (a) oil production reduction and (b) no reduction (i.e., oil production continu-ation), and a set of the profit pair ($xxx/$yyy) describes the profit of Kuwait first, and that of Saudi Arabia in the latter. As mentioned above, in Cell A, both decide on "oil reduction," then both earn. But in Cell C the oil price decrease is moderate (because Kuwait's production amount is not large), and Kuwait wins largely (this is the motivation for Kuwait's betrayal), but Saudi loses. In contrast, the betrayal (i.e., no oil production reduction) of Saudi Arabia is very serious. In Cell B, when Saudi Arabia continues oil production, the oil price decrease around the world is significant, leading to huge losses for Kuwait if Kuwait tries to keep to the cartel agreement. Note that Saudi Arabia enjoys a small profit due to the Kuwait's share loss. If both players renege in Cell D (i.e., both countries do not reduce oil production), both suffer big losses.

1.1.2 *Decision-Making*

Supposing that the reader can roughly understand the above *economic analysis*, now, my question to the reader is on politics or decision-making: "Suppose you

Table 1.1.   Profit & deficit for Kuwait & Saudi Arabia according to the oil production strategy.

	Saudi Arabia	
	Production Reduction	No Reduction
Kuwait Production Reduction	A: $300M/$100M	B: –$800M/$200M
No Reduction	C: $500M/–$1000M	D: –300M/–$2000M

are an advisor to the Minister of Energy, Industry & Mineral Resources, Saudi Arabia, how will you advise the Minister *to decide on the action to take from a political viewpoint?*" Below, I propose three types of advice, which were reasonable in the 1970s.

(1) Economic viewpoint

When Kuwait opts for "Cartel Cooperation (i.e., Reduction)," it is more beneficial for Saudi Arabia to take the "No Reduction" option; however, when Kuwait takes the route of "No Reduction," it is of greater benefit for Saudi Arabia to adopt the "Reduction" approach. The more beneficial decision for Saudi Arabia changes according to Kuwait's choice of action; in other words, there is no "Dominant Strategy" for Saudi Arabia. However, from the viewpoint of Kuwait's Minister of Oil, regardless of Saudi Arabia's strategy, "No Reduction" is their *"Dominant Strategy,"* i.e., when Saudi Arabia adopts "Reduction," Kuwait's profit in Cell C ($500M) is larger than that in Cell A ($300M), and when Saudi Arabia takes "No Reduction" Kuwait's profit in Cell D (–$300M) is better than that in Cell B (–$800M). Taking into account that Kuwait's dominant strategy is "No Reduction," the most beneficial strategy for Saudi Arabia would be to go for "Production Reduction" (Cell C). This is the straight-forward theoretical conclusion based on the payoff table in Table 1.1 under the supposition that both players (Kuwait and Saudi Arabia) are rational, and by taking into account the economic/financial logical viewpoint, which is close to *Liberalism.*

(2) Power Balance Viewpoint

Sometimes the decision-making process depends on individual leaders' personalities (*actor-specific idiosyncrasy*). In the event that Kuwait betrays the cartel, Saudi Arabia may consider evoking the *Penalty* in the OPEC regime to keep the international power balance as OPEC leader; that is, by taking "No Reduction," Saudi Arabia causes to Kuwait to suffer a large deficit (–$300M) in Cell D, by taking on some self-sacrificing deficit. Though the sacrificial deficit for Saudi Arabia is huge (–$2000M), Saudi Arabia may take this action if they approach from the *Realism* viewpoint. Remember the *Almond–Lipmann consensus,* which states that public opinion is volatile and irrational, and thus a dubious basis for foreign policy, and the political leader may not seriously consider the opinion of the general public. How one integrates this sort of actor-specific idiosyncrasy into the payoff table is discussed in Chapter 8.

(3) Psychological/Cultural Milieux

If Saudi Arabia's Minister is a *Constructivist* who places respecting *worldwide norms* as the highest priority, however, Cell D (No Reduction) would not be a recommended course of action because both countries' continuous oil production will cause a serious oil price decrease that would likely lead to worldwide economic recession. The most recommended action would be Cell C. Note that though the decision is the same as in case (1), the background reason is completely different.

While there may not be any unique actions to recommend as an advisor to the Saudi Arabia Minister based on the theories presented above, the actual action taken at the end of the day may also be determined by the political leader's personality or character, and differ from what is advised. Note also that the options that a government advisor may recommend to Saudi Arabia's Minister also depend on the nation's psychological and cultural milieu [Chapter 8]. The three factors (*profit, power, and norm*) that affect political decision-making are discussed in Chapter 3.

1.1.3 *Influence of Advanced Technology*

I would like to emphasize that *advanced technology*, such as the development of hydraulic fracturing (fracking) on *Shale Oil/Gas mines*, has dramatically changed the Oil-Economic Power Balance in the world [Chapters 4–6], i.e., offering completely different factors for analysis from those considered in the above three types of advice. This change was what motivated me to write this book. In 2014, OPEC, led by Saudi Arabia, initiated an *"economic oil war"* against the United States. Saudi Arabia attempted to cut oil production in order to keep oil prices high, as it had previously done, but the US countered by seeking to increase its shale oil/gas production. In retaliation, OPEC attempted to drive some US *shale oil producers into bankruptcy* and reduce the flow of North-American shale oil into the global market by increasing oil production in November — a move which drove oil prices down to nearly US$50/barrel; the price at which many US shale producers could not even *break-even* [see section on risk management in Chapter 7 for more information on this]. However, this strategy did not work out the way OPEC wanted; the deficit OPEC incurred was significantly greater than the US's. In short, the situation demonstrated that oil prices could no longer be solely controlled by OPEC; and that control has gradually shifted to the United States recently (i.e., the old fashion oil cartel is disappearing). We now need to change the game model used in our analysis from the two players/actors model to the three (or more) players/actors model.

Lesson from this Oil War: While over half of the proven oil reserves are generally under the control of OPEC, there are many more unconventional reserves, such as oil shale, heavy oils and tar sands, outside the Middle East (see Chapter 6). And most of these are on the edge of affordability. The key to unlocking these reserves, is the development of advanced technology that can produce oil at an affordable price; thus leading to a dramatic change in the international political power balance. Supply from the short-cycle US oil market is required nowadays to balance the global crude oil market at a rate where US shale should remain a growth industry (~US$70/barrel). In other words, in order to increase or at least maintain USA's power, advanced key technology development strongly initiated by politics and policymaking (i.e., "Politico-Engineering" or "politically initiated Engineering" — the subject of this book) is needed.

1.2 Four Influencing Factors on Engineering

In Chapter 1, we discuss the historical background of "Politico-Engineering" (i.e., politically-initiated engineering). There are primarily four factors that have influences on engineering: (1) Society/culture/religion, (2) Technology/science, (3) Economics, and (4) Politics/law — henceforth known as "STEP". Although MBA classes normally teach the acronym "PEST", I propose using *STEP* instead; taking into account their chronological sequence in modern history (see Fig. 1.1).

The strength of the impact of these factors was different according to history. Alchemy in the 16th century is an example of *Socio-Engineering*. At that

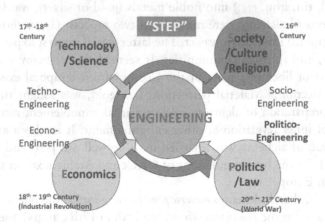

Fig. 1.1. Historical aspect of the four factors which influences on Engineering: Socio-, Techno-, Econo-, and Politico-engineering.

time, the "Copernican (Heliocentric) model" was denied by the church (as per the accepted doctrine of the time), but "alchemy" was approved. Religion was controlling science. In the 17th and 18th centuries, engineering grew to become driven by Science and Technology (S&T); the so-called *Techno-Engineering* move was respected then, in place of the dominant influence of the Church in the 16th century. In the 18th and 19th centuries, technologies for mass production at low manufacturing cost were required and *Econo-Engineering* became the mainstream; i.e., technologies that are essential for enhancing national strength (economically and politically, both domestically and abroad). The 19th and 20th century desires to increase national wealth and military strength increased friction and led to the First and Second World Wars in the 20th century. Engineering during this period mainly manifested in the form of government-led production of weapons of warfare — marking the beginning of the era of *Politico-Engineering*. The growth of "Politico-Engineering" after the Wars will be discussed in Chapter 2.

1.3 Socio-Engineering

This section is highly in debt to two books authored by Gerald Holton [1] and Arnold B. Arons [2]. These books are very useful for learning the detailed history of social engineering (or socio-engineering).

1.3.1 *Alchemy*

In the 16th century, *alchemy*, the technology that turned base metals (copper, lead, tin, iron, etc.) into noble metals (gold or silver), was very popular. Discussions of alchemy were made from two aspects: (1) exoteric practical application and (2) the esoteric aspect. The latter was of interest to psychologists, philosophers, and spiritual communities. It seemed that alchemy was believed to be "the art of liberating parts of the Cosmos from temporal existence and achieving perfection." Material perfection, i.e., gold, was sought through the action of a preparation of alchemy, while spiritual ennoblement resulted from some form of inner revelation or other enlightenment. It has been argued that this perception of alchemy was what led it to be well accepted and supported by the church by the time translations of Greek and Arabic texts on the subject reached Latin Europe.

On the other hand, alchemy's *exoteric practical application* was pursued by physical scientists. During the *Renaissance* (from the 14th to 17th century), the European alchemy community — birthed at the dawn of medical and pharmaceutical

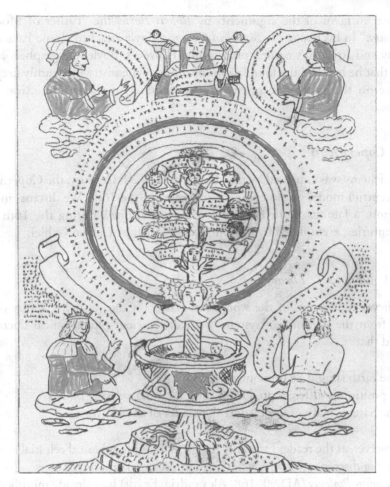

Fig. 1.2.   Alchemic treatise of Ramon Llull, 16th century.

sciences — that had been perceived as a community dedicated to the occult, was restored. Figure 1.2 illustrates the Alchemic treatise of *Ramon Llull*, used in the 16th century. Alchemy was taught at universities during that era as a regular course, in parallel to Paracelsus (the natural healing of a magical world, based on alchemy) and Aristotle's philosophy of nature.

The reader may remember the name, *Robert Boyle* (1627–1691), whose law is usually learned in high school chemistry, i.e., that pressure exerted by a given mass of an ideal gas is inversely proportional to the volume it occupies, namely pV = constant, if the temperature and amount of gas remain unchanged within a closed system. It was during the 17th century that practical alchemy started to disappear in favor of the arguments by *Robert Boyle*, the "Father of Modern Chemistry." In his book, *The Skeptical Chymist*, Boyle attacked classic Paracelsus alchemy and Aristotle's natural philosophy. However, Boyle's biographers mentioned that he had laid the foundations of modern chemistry by steadily keeping his base on the scholastic sciences and alchemy in theory, practice, and doctrine.

1.3.2 *Copernican*

While alchemy was approved by the church in the 12th Century, the Copernican (Heliocentric) model would come to be denied in the 16th. We discuss, in this subsection, a famous *Battle between Science and Religion* during the 16th and 17th centuries, exemplified by the "inquisition" against Galileo Galilei.

1.3.2.1 *Geocentric Theory*

A medieval conception of the world was based on the *geocentric theory* originating from the Aristotelian system. Basically, ancient philosophical doctrines believed that:

(1) The earth should be immobile.
(2) Its position is differentiated from those of other celestial bodies.
(3) The natural place of the earth is the center of the Universe.

However, as the reader can easily imagine, a simple spherical celestial motion induces big discrepancies with what astronomic observations tell us.

Claudius Ptolemy (AD 90–168, Alexandria, Egypt) had already modified the original pure-circle model into the model shown in Fig. 1.3(a), by adjusting the planets' respective axes, direction of motions, rates and radii of rotations, the number and size of epicycles, and eccentrics. He invented a special astronomic apparatus called the *Equant*, which could simulate the planetary motion depicted in Fig. 1.3(a). This proved very useful to astronomers, navigators, and astrologers more than 14 centuries later, until the Heliocentric model was accepted.

The key person who connected Geocentric theory to Christian doctrine is *Thomas Aquinas* (1225–1274), a Catholic priest and Dominican friar in Italy. Thomas embraced several ideas proposed by *Aristotle* — whom he referred to as "the Philosopher" — and attempted to synthesize Aristotelian philosophy, including Geocentric theory, with the principles of Christianity. Thomas was

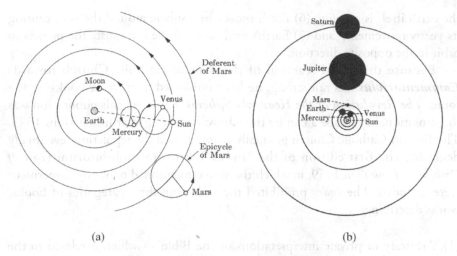

Fig. 1.3.   (a) Geocentric model — Partial and schematic diagram of Ptolemaic system of planetary motion, and (b) Heliocentric model.

honored as a Saint by the Catholic Church and is held to be the model teacher (*Doctor of the Church*) for those studying for the priesthood. Accordingly, Geocentric theory became a part of Christian theology — after which came the so-called *Dark Ages* in science and philosophy, which continued until the 16th century. At the time, questioning the central Church and its doctrine was prohibited, and no one had appeared to attempt battle against such forbidden areas.

1.3.2.2 *Heliocentric Theory*

Nicolaus Copernicus (1473–1543) was a Renaissance mathematician and astronomer who, despite obtaining a doctorate in canon law and having officially succeeded to the Warmia canonry in 1497, developed his own celestial model of a *heliocentric planetary system* (Fig. 1.3(b)). Because he worried about facing accusations of heresy from the Roman Catholic Church, Copernicus initially only distributed handwritten copies of the book known as the **Commentariolus** to a few select friends around 1514. It was in the **Commentariolus** that Copernicus set out his view of the universe. He described seven points which support the heliocentric model of the solar system: (1) Planets don't revolve around one fixed point; (2) the earth is at the center of the moon's orbit; (3) The sun is at the center of the universe, and all celestial bodies rotate around it; (4) The distance between the earth and the sun is only a tiny fraction of the distance of other stars from the earth and Sun; (5) Stars do not move, and if they appear to, it is only because the earth itself is moving; (6) Earth moves in a sphere around the sun, causing its yearly movement; and (7) Earth's orbit around the sun causes the planets to orbit in the opposite direction.

Because the initial reaction of the Roman Catholic Church towards **Commentariolus** was not strong, he later published a set of six books on the topic: **The Revolution of the Heavenly Spheres**. However, this publication was too sensational to escape a ban by the Church not long after his death in 1543. The Roman Catholic Church gradually strengthened the regulation, eventually declaring the first edition of the "Index Librorum Prohibitorum (*List of Prohibited Books*)" in 1559, in which the above-mentioned books by Copernicus were included. The *Index* prohibited the following three categories of books; works describing:

(1) Arbitrary or private interpretations of the Bible — which is related to the Church's attack on *Protestantism* and the German Catholic priest, *Martin Luther* (1483–1546).
(2) Theories regarding the Cosmos or Nature differing from the model by *Thomas Aquinas* (i.e., Aristotle's philosophy of nature, that was accepted by the Christian doctrine of the time) — which is the origin of the battle between science and religion.
(3) Neo-Platonism and its related *Mysticism*, i.e., Magic — which accelerated the so-called "Witch hunts."

Although Copernicus had built his own modest observatory in 1513 so that he could methodologically observe the planets at any time, his observations did lead him to form inaccurate conclusions — including his assumption that the orbits of planets occurred in perfect circles. Copernicus' error was 'rectified' by his 'successor', German astronomer Johannes Kepler; who proved later in the 17th century that planetary orbital paths were actually oval in shape.

1.3.3 *Battle between Science and Religion*

Johannes Kepler, Galileo Galilei, and Isaac Newton — three famous scientists who triggered the "techno-engineering" period, are described in this subsection.

Johannes Kepler (1571–1630) is well known as a German mathematician and astronomer. But during his lifetime, he was more famous as an astrologer and mysticist. His main work was to establish the theoretical principles of astrology, which had a corresponding practical part that dealt with the making of annual astrological forecasts about individuals, cities, the human body, and the weather.

Though he was a devotee of a Protestant sect under Martin Luther, his heliocentric theory — for which he was later excommunicated — was not accepted by the Protestant Church. Compared to Galileo, Kepler was lucky to have merely been *excommunicated* for his theory rather than sentenced to be burnt at the stake (a punishment typically issued by the Inquisition). However, while he himself escaped the stake, his mother was, unfortunately, accused of being a witch and arrested in 1619 when Kepler publicized his heliocentric planetary law.

The most miserable victim was *Galileo Galilei* (1564–1642), who was found guilty of heresy by the *Roman Inquisition* in 1633 because of his strong support of the Copernican theory. Galileo remained under house arrest until he passed away nine years later at the age of 77. Pope Urban VIII (through the *Roman Curia*) did not allow anyone to hold a funeral for him.

The inquisition against Galileo was concocted based on forged documents, and the sentence is now believed to be a misjudgment by the Vatican (Fig. 1.4) [3]. The reason Galileo was found guilty, was primarily due to Galileo publishing his *Dialogue on the Two Chief World Systems* — a work that basically supported the heliocentric theory, which the Church regarded as "contrary to Holy Scripture." This book had actually initially been published in Florence, in 1632, under a formal license from the Inquisition, after abiding by extensive

Fig. 1.4.   The inquisition against Galileo Galilei (1564–1642) and the Heliocentric model. [Based on Cristiano Banti's 1857 painting]

corrections suggested by the Inquisition. However, because of unexpectedly significant reactions to this publication, in addition to the Dominicans, who had accused Galileo of heresy from the beginning, the Society of Jesus also jointly pushed for the Roman Catholic Church to list the ***Dialogue*** in the Prohibited Books Index six months after its publication. Accordingly, Galileo was called to the Inquisition based on an alleged old "warning" document issued in 1616 by the Inquisition to cease teaching the Copernican theory. It was through later investigations that this document — without a signature by the Cardinal, any notarization or the Seal — was revealed to (possibly) be a fake obtained by the opponents of Galileo by bribing the Inquisition secretary. Anyhow, at that time, once an inquisition was called, most of the cases resulted in a "guilty" verdict, particularly because severe tortures were often used to obtain a confession prepared by the inquest. According to popular legend, after his famous *Abjuration*, Galileo allegedly muttered "yet it [the earth] moves [rotates]" (*Eppur si muove*).

Because the key document, the "warning" allegedly issued by the Roman Catholic Inquisition in 1616, was discovered to be a fake, the guilty sentence against Galileo was cancelled in 1757 — more than 120 years later. His book ***The Dialogue on the Two Chief Systems of the World*** was also delisted from the Index of prohibited books in 1835 — which was great for science, but too late for Galileo Galilei.

The reader may know *Isaac Newton* (1643–1727) as a "rationalist" who developed modern scientific philosophy (Fig. 1.5). Since Newton's law of motion and "gravitational law" automatically support the heliocentric model, he could have been summoned to an inquisition in the worst case scenario. Thanks

Fig. 1.5.   Isaac Newton (1643–1727).

to his reputation as an alchemist, Newton did not have any problems with the Roman Inquisition.

When Newton published his work, *Principia*, summarizing what we know today as Newton's Law of Motion (which is considered one of the triggers of modern science), a half century had passed since Galileo's inquisition. In the original preface to his work, Newton described:

> *"Since the ancients esteemed the science of mechanics of greatest importance in the investigation of natural things, and the moderns, rejecting substantial forms and occult qualities, I have in this treatise cultivated mathematics as far as it related to philosophy ... the whole burden of philosophy seems to me to consist ... from the phenomena of motions to investigate the forces of nature, and then from these forces to demonstrate the other phenomena ... in the Third Book I derive from the celestial phenomena the force of gravity with which bodies tend to the sun and the several planets. Then from the forces, by other propositions which are also mathematical, I deduce the motions of the planets, the comets, the moon, and the sea ..."*

Newton is considered a key person of modern science because of his rejection of "substantial forms and occult qualities." In parallel, the reader can understand that he challenged traditional Roman Catholic doctrine. Newton was the first scientist who was knighted by Queen Anne of England in 1705. His strong interest in religion was shown in his various contributing papers and activities against traditional Catholic doctrines. Because the English Royal Family was revolting against the Roman Catholic Church at that time, Newton's activities were well protected in England — a large contrast to what Galileo Galilei faced.

Interestingly, after publishing *Principia* at the age of 45, Newton's major work shifted to focus on alchemy for the next 40 years, until his demise. As I explained earlier, alchemy was an established traditional scientific territory in the 17th and 18th centuries. Alchemists respected "spiritual ennoblement" because they believed that the alchemist's spirit significantly affected the materials' transformation processes. Thus, they practiced *abstinence* daily, and *meditation* and *devotion* prior to the experiment (e.g., praying for the alchemic processes' success) were popular. Newton seemed to be an alchemist who continuously practiced meditation. Like various scientists in the "New-Science" sect, Newton's discovery on the law of motion may have come about during one such meditation. Newton once wrote:

> *"And the same year (at age 24) I began to think of gravity extending to the orb of the Moon, and ... from Kepler's Rule ... I deduced that the forces which keep the Planets in their orbs must be reciprocally the squares of their distances from the centers about which they revolve: ... All this was in the two plague years of 1665 and 1666. In those days I was in the prime of my age for invention, and minded*

Mathematics and Philosophy more than at any time since." [*The Great Plague* *("Black Death") was the last major epidemic of the bubonic plague due to pests to occur in the Kingdom of England.*]

The above description suggests that Newton was very religious and his discovery came from a sort of "revelation" during his meditation. In the days of the Socio-Engineering era, "religion" controlled "science".

1.4 Techno-Engineering

In the 17th and 18th centuries, religion gradually lost its monopoly, and engineering based on pure Science & Technology (i.e., *Techno-Engineering*) became respected instead. This period is also known as the *Age of Reason* and Age of Enlightenment. The period in which Isaac Newton (1643–1727) published his law of universal gravitation seemed to be the dawn of scientific freedom; it was then that Benjamin Franklin (1706–1790) clarified the origin of electricity in the (then) developing nation of America. As the reader probably knows (see Fig. 1.6), Franklin is said to have used a child's kite to prove that lightning is really a stream

Fig. 1.6.   Benjamin Franklin (1752) flying a kite in a thunder storm. Based on the drawing by Bernard Hoffmann.

of electrified air (a reckless challenge, indeed!). Mathematics and pure physics drastically progressed during this period.

1.4.1 *Newton and Descartes*

Many people believe that modern scientific thought was established by Descartes and Newton. *René Descartes* (1596–1650) was a creative French mathematician, an important scientific thinker, and an original metaphysician during the early 17th century — almost the same period as Galileo Galilei.

As Newton clearly declared, the law of motion and the law of universal gravitation were not solely his own discoveries, but that they had arisen from many of the concepts (e.g., force, acceleration, vector quantities, and the first law of motion) that came to him from Kepler, Galileo and their followers. The law of universal gravitation was also deduced from a variety of sources such as *Tycho Brahe*'s (1546–1601) observations of the moon's motion. So, what is the actual contribution by Newton? He successfully combined the principally inductive approach with the deductive method most prominently displayed by Descartes. With his mathematical powers enriching the experimental attitude, Newton set a clear, straight course for the methods of physical science.

"I think, therefore I am," is a famous philosophical proposition by *René Descartes*. The simple meaning is that thinking about one's existence proves that "I" exists to do the thinking; or, as Descartes explains, "we cannot doubt of our existence while we doubt …" This proposition became a fundamental element of Western philosophy, as it was perceived to form a foundation for all knowledge. In his publication in 1644 titled ***Principles of Philosophy***, Descartes described:

> *"While we thus reject all of which we can entertain the smallest doubt, and even imagine that it is false, we easily indeed suppose that there is neither God, nor sky, nor bodies, and that we ourselves even have neither hands nor feet, nor, finally, a body; but we cannot in the same way suppose that we are not while we doubt of the truth of these things; for there is a repugnance in conceiving that what thinks does not exist at the very time when it thinks. Accordingly, the knowledge, "I think, therefore I am", is the first and most certain that occurs to one who philosophizes orderly."*

Descartes suggested a philosophy now known as *"mind–body dualism"*. It argues that the mind and the body are not identical; that the body works like a machine and has material properties, while the mind (or soul), on the other hand, is non-material and does not follow the laws of nature. This form of dualism or duality proposes that the mind controls the body, but that the body can also influence the otherwise rational mind; such as when people act out of passion.

Descartes' "dualism" drastically changed the situation for the "battle between science and religion" by treating "mind/spirit" and "body/material" separately. The Cosmos began to be considered a huge mechanical system composed of various materials, such that its structure would not provide significant problems to the religious spirit. This philosophy was also adopted in Western medicine; where the *human body is considered a mechanical system* composed of various organs, and a disease as a failure of a mechanical part. It was through the adoption of this attitude that surgery became popular in the Western medical community; as an act of repairing the "mechanical part" of a person. This point significantly differs from the beliefs of Eastern medicine, in which the healing of the body is directly combined with the healing of the mind and spirit. Interestingly, techniques from Eastern medicine for the treatment of conditions ranging from back pain to depression, and even cancer, have recently spread over to the Western community.

1.4.2 *Boyle*

As already introduced in the previous section, *Robert Boyle* (1627–1691) is known as the "Father of Modern Chemistry." Boyle's first study entitled **New Experiments, Physico-Mechanical, Touching the Spring of the Air and Its Effects** (1660) was pneumatic. He continued his study of air and vacuum throughout the rest of his life, and established the famous Boyle's law. Although his experiments with the "Boyleian vacuum" were repeated by many, no one in the 17th century surpassed Boyle's ingenuity or technique.

Boyle's law was extended by his assistant, *Robert Hooke* (1635–1703) to derive Hooke's Law, a law determining general elastic performance. Boyle's law (pV = constant) could deduce a model in which the pressure comes from the averaged repulsive force from each gas particle. Boyle suggested that the Nature of Heat "consists" in "various, vehement and intestine commotion of the parts among themselves," from which Hooke proposed that "Heat is a property of a body arising from the motion or agitation of its parts." These led to the discovery of another law later known as Charles' law (by *Jacques A. C. Charles*, 1746–1823), i.e., that the pressure exerted by a given mass of an ideal gas is directly proportional to the temperature (we need to use the absolute temperature, °K), namely p = constant $\times$ T.

We now have Boyle–Charles' Law, which is a must in high-school chemistry: $pV = nRT$, where p is the gas pressure, V volume, n the number of moles of a fixed mass of gas, and R is the gas constant (R = 8.314 J/mol·K). Many 17th century thinkers and scientists held a "corpuscular" view of the structure of matter, and at least some of these had established a clear association between heat and the motion of the constituent particles.

1.4.3 *New Mathematics*

During the Renaissance (17th and 18th centuries), mathematics and scientific ideas experienced a sort of explosion across Europe, free from religious constraints. Thus, this period is sometimes called the *Age of Reason*. As discussed in the previous section, the "Copernican Revolution" in the 16th century led to the rise (and unfortunate falls) of scientists like *Galileo Galilei, Giordano Bruno* (1548–1600), *Michael Servetus* (1511–1553), and *Johannes Kepler*, leading finally to Kepler's formulation of the mathematical laws of planetary motion.

The 17th century began with the invention of the *logarithm* by *John Napier* (1550–1617). The logarithm contributed to the advance of science, in particular astronomy, by making some difficult calculations relatively easier. Without "log," Kepler and Newton could never have performed the complex calculations needed for their innovations. In 1637, *René Descartes* published his ground-breaking philosophical and mathematical treatise, **Discourse on Method**, and one of its appendices, **La Géométrie** ("The Geometry"); events now considered landmarks in the history of mathematics. **The Geometry** introduced what has become known as *standard algebraic notation*, using lowercase *a*, *b*, and *c* for known quantities, and *x*, *y*, and *z* for unknown quantities. Descartes' work, referred to as analytic geometry or Cartesian geometry, had the effect of allowing the *conversion of geometry into algebra*. As a point moves along a curve, then, its coordinates change; but an equation can be written to describe the change in the value of the coordinates at any point in the figure. Using this novel approach, you can understand that an equation, for example, $x^2 + y^2 = 1$ describes a circle; $y = ax^2$ a parabola; $x^2/a^2 + y^2/b^2 = 1$ an ellipse; and $x^2/a^2 - y^2/b^2 = 1$ a hyperbola, which accelerated the cosmic model analyses. *Blaise Pascal* (1623–1662) is well known as a discoverer of a physical law named after him: that pressure exerted anywhere in a confined liquid is transmitted equally and undiminished in all directions throughout the liquid. If you are a scientist or an engineer, you must have been trained to use the international unit for the measurement of pressure, Pa, which is equivalent to N/m^2. Here, you use "Pascal" and "Newton" for pressure and force units.

Isaac Newton is called here again. His work, **Philosophiae Naturalis Principia Mathematica** ("Principia") is considered to be among the most influential books in the history of science, and it dominated the scientific view of the physical universe for the next three centuries. He presented his theories of motion with a detailed introduction of the concepts of *differentiation* and *integration*. Historically, *Gottfried Leibniz* (1646–1716) developed the same idea independently, and had published it in 1684 — slightly earlier than Newton's **Principia** (published in 1687).

A derivative function $f(x)$ (denoted as $f'(x)$ or $df(x)/dx$) provides the slope at any point of a function $f(x)$. This process of calculating the slope or derivative of a curve or function is called differential calculus or differentiation. For example, $f(x) = -ax^2 + bx + c$ is a popular parabolic curve when the object is thrown under a gravitational atmosphere. As the reader can easily solve, the moving direction of the object (i.e., slope) is calculated as $(-2ax + b)$, and the object peak point is obtained at $x = b/2a$. In contrast, the integral of a curve can be thought of as the formula for calculating the area bounded by the curve and the x axis between two defined boundaries. For example, the integrals of $f(x) = x^r$ (power), $1/x$ ($r = -1$) and ex are $x^{r+1}/(r+1)$, $\ln(x)$ (logarithmic function) and e^x, respectively, which are some of the most useful formulae in physics.

Bernoulli's family produced multiple mathematicians over a couple of generations at the end of 17th and the beginning of 18th century, such as his brothers *Jacob Bernoulli* (1654–1705) and *Johann Bernoulli* (1667–1748). One well-known and topical problem was that of designing a sloping ramp, which would allow a ball to roll from the top to the bottom in the fastest possible time. *Johann Bernoulli* demonstrated through calculus that neither a straight ramp nor a curved ramp with a very steep initial slope were optimal, but that actually a less steep, curved ramp, known as an "inverted cycloid" curve (a kind of upside down cycloid; similar to the path followed by a point on a moving bicycle wheel) is the curve of fastest descent. In fluid dynamics, *Bernoulli's Principle* states that for an inviscid flow of a non-conducting fluid, an increase in the speed of the fluid occurs simultaneously with a decrease in pressure or a decrease in the fluid's potential energy. The principle is named after *Daniel Bernoulli* who published it in his book, **Hydrodynamica**, in 1738.

Leonhard Euler (1707–1783), one of the Bernoulli's young students, was a great Swiss mathematician, covering a wide range of mathematics; from geometry to trigonometry. He even managed to combine several of these together to produce one of the most beautiful of all mathematical equations, $e^{j\pi} = -1$, sometimes known as Euler's Identity. This equation combines arithmetic, calculus, trigonometry, and complex analysis into what has been called "the most remarkable formula in mathematics," and has been said to be "filled with cosmic beauty." Another such discovery, often known simply as *Euler's Formula*, is $e^{jx} = \cos x + j \sin x$. In fact, in a recent poll of mathematicians, three of the top five most beautiful formulae of all time were Euler's. He seemed to have an instinctive ability to demonstrate the deep relationships between trigonometry, exponentials, and complex numbers.

1.4.4 *Franklin*

After the verification of magnets by *William Gilbert* (1540–1603), the following 200 years from 1600 to 1800 were the *electrostatic* age, focusing primarily on the electrostatic phenomena and the nature of electricity. *Benjamin Franklin* (1706–1790), sixth President of Pennsylvania and one of the founding members of the United States, was "the First American" in many ways. In Pennsylvania State stands the Ben Franklin Foundation, which honors his significant scientific and political contributions to the state. Ben Franklin Technology Partners is one of the nation's longest-running technology-based economic development programs, which has provided both early-stage and established companies with funding, business and technical expertise, and access to a network of innovative, expert resources.

Franklin's discovery came from his investigation of electricity. In 1750, he published a proposal for an experiment to prove that lightning is electricity, by flying a kite in a storm that appeared capable of becoming a lightning storm. On June 15, Franklin conducted his famous kite experiment in Philadelphia, successfully extracting sparks from a cloud to be trapped in a Leyden jar (a capacitor) (Fig. 1.6). Note that multiple scientists, such as *Georg Wilhelm Richmann* (Russia, 1711–1753), were indeed electrocuted during the months following Franklin's experiment.

Franklin indicated in his publication that he was aware of the danger and offered alternative ways to demonstrate that lightning was electrical, as shown by his use of the concept of electrical ground. If Franklin did perform this experiment, he may not have done it in the way that is often described — flying the kite and waiting to be struck by lightning — as it would have been dangerous. Instead he used the kite to collect some electric charge from a storm cloud, which implied that lightning was electrical. On October 19, in a letter to England with directions for repeating the experiment, Franklin wrote:

> "*When rain has wet the kite twine so that it can conduct the electric fire freely, you will find it streams out plentifully from the key at the approach of your knuckle, and with this key a phial, or Leiden jar, may be charged: and from electric fire thus obtained spirits may be kindled, and all other electric experiments performed which are usually done by the help of a rubber glass globe or tube; and therefore the sameness of the electrical matter with that of lightening completely demonstrated.*"

Franklin's electrical experiments led to his invention of the *lightning rod*. He noted that conductors with a sharp rather than smooth point could discharge silently, and at a far greater distance. He supposed that this could help protect buildings from lightning:

> "[by attaching] *on the highest parts of [...] edifices, upright rods of iron, made sharp as a needle and gilt to prevent rusting, and from the foot of those rods*

a wire down the outside of the building into the ground [...] would not these pointed rods probably draw the electrical fire silently out of a cloud before it came nigh enough to strike, and thereby secure us from that most sudden and terrible mischief?"

1.4.5 *Coulomb*

Charles Augustin Coulomb (1736–1806, Fig. 1.7) established quantitative treatments in electricity quantity, electrostatic force, and charge capacity. He invented a torsional force measuring balance with a metal ball hung on a thin wire for measuring the micro-force between two electric charges or two magnetic charges (1777). According to him:

> *"... the moment of the torque is, for wires of the same metal, proportional to the torsional angle, the fourth power of the diameter and the inverse of the length of the wire ..."*

His detailed measurement verified that the electric attractive or repulsive force is inversely proportional to the square of distance r, i.e., $f \propto q_1 q_2 / r^2$. This formula derived that both electric and magnetic charges should be a scholarly quantity, analogous to the mass in Newton's law of universal gravitation. Coulomb also concluded that there are two kinds of charges (+ and −), as opposed to Franklin's argument (that there was only one type). It was for this achievement that Coulomb is considered the pioneer of static electromagnetic theory.

Fig. 1.7.   Charles-Augustin de Coulomb (1736–1806).

1.4.6 *Huygens*

Christiaan Huygens (1629–1695) was a prominent Dutch mathematician and scientist. He is known particularly as an astronomer, physicist, and horologist. His work included early telescopic studies of the rings of Saturn and the discovery of its moon, Titan; the invention of the pendulum clock; and other investigations in timekeeping. Huygens is remembered especially for his *wave theory of light*, which was published in 1690 in his **Treatise on light**. He refers to *Ignace-Gaston Pardies*, whose manuscript on optics helped him on his wave theory.

A basic principle of Huygens' is that the speed of light is finite; a point which had been the subject of an experimental demonstration by *Olaus Roemer* in 1679, at the Paris Observatory. The theory is what we know today as kinematics and its scope is largely restricted to geometric optics. It deals with wave fronts and their normal rays, with propagation conceived by means of spherical waves emitted along the wave front. It was justified as an "ether" theory, involving transmission via perfectly elastic particles — a revision of the view by Descartes. The nature of light was therefore defined as a longitudinal wave.

In 1672, Huygens had experimented with birefringence (i.e., double refraction) in Icelandic spar (calcite); a phenomenon discovered in 1669 by *Rasmus Bartholin*. He explained it with his wave front theory and concept of evolutes (envelope curves of the normals). He also developed ideas on caustics. Newton, in **Opticks** (published in 1704), had instead proposed a *corpuscular theory of light*. Huygens' theory was not accepted because longitudinal waves cannot show birefringence. It was only in 1801 that *Thomas Young* proved Huygens' wave theory through interference experiments, whose results could not be explained with light particles. The solution to the problem Huygens had faced was then resolved by a *transverse wave theory*. Anyhow, this contradiction was explained later by modern physics' *wave–particle duality*.

1.4.7 *Discovery of Piezoelectricity*

The dawn of piezoelectricity actually happened at the end of 19th century (in the midst of the Econo-Engineering period). However, for the sake of this story, its history has been inserted into this subsection.

The Curie brothers, *Pierre Curie* (1859–1906, Fig. 1.8) and *Paul-Jacques Curie* (1856–1941), discovered the *direct piezoelectric effect* in single crystal quartz in 1880. Under pressure, quartz generated an electrical charge/voltage. The root of the word "piezo" means "pressure" in Greek; hence the original meaning of the word piezoelectricity implied "pressure electricity." Materials showing this phenomenon conversely also have a geometric strain proportional to an applied electric field. This is the *converse piezoelectric effect*, discovered by

Fig. 1.8.   Pierre Curie (1859–1906).

Gabriel Lippmann in 1881. Recognizing the connection between the two phenomena helped *Pierre Curie* to develop pioneering ideas about the fundamental role of symmetry in the laws of physics. We needed to wait more than 30 years until practical application developments using piezoelectricity (e.g., underwater sonar systems) started.

1.5 Econo-Engineering

This section is in debt to a book authored by Shuzo Miwa [4]. In the 18th and 19th centuries during the *Industrial Revolution*, technologies for mass production at low manufacturing cost were required, and *Econo-Engineering* (the engineering of technologies that are essential to enhance national strength) became the mainstream. For example, in the early 18th century, "British textile manufacturing" meant the processing of wool by individual artisans, who did the spinning and weaving on their own premises. The roller spinning machine developed for drawing wool to a more even thickness increased overall productivity. A steam engine was commercially successful in the application for pumping water out of mines. The profits generated by West European counties during the industrial revolution were directly invested into *colonization* and country-strengthening strategies prior to World Wars I and II.

1.5.1 *Industrial Revolution*

England expanded its international trade dramatically in the 18th century, based on the theory and practice of *mercantilism*. Both the development of the military and the expanding international trade that grew from mercantilism needed advanced technologies to manufacture steel and ships. Because both industries required significant amounts of wood as fuel, trees were indiscriminately cut down and harvested, leading to a decline in available wood-fuel. Consequently, an alternative fuel was needed, and the demand for coal increased as a result (i.e., *charcoal-to-coal shift* of iron manufacturing). Concurrently, mining tunnels were penetrating deeper and deeper underground, causing the need for proper drainage and air circulation to become major problems in mining. And because the primary power sources in the 18th century were horses and watermills, when the cost of fodder inflated during the wars, this cost made-up up to three-quarters of the total coal mining cost. This need accelerated the development of new power supplies/machines.

On the other hand, we should also point out the drastic improvement in life-style from 1750–1850 for both aristocrats and regular people. At that time, oriental products, exemplified by high-quality Chinese and Japanese Art and ceramic ware with sophisticated designs, and beautifully dyed Indian cotton cloth, were highly sought after by most European noblemen; and these were largely imported. However, the typical materials exported from Europe, e.g., woolen fabric, were not respected in Asia at all; leading to the unidirectional flow of money from Europe to Asia. In order to recover from this economic imbalance, the industrial revolution became the highest priority in England, particularly in improving textile/woolen fabric manufacturing capabilities, so as to reduce the price of these products, in order to compete with Indian cotton cloths. In other words, we can say that one of the motivations for the *Industrial Revolution* in Europe came from a sort of cultural complex of Europeans against Oriental goods [4].

The main achievement of the Industrial Revolution was a "*mechanics revolution*", exemplified by spinning and weaving machines, steam engines/turbines, and machine tools, which we review in the following sub-sections.

1.5.2 *Spinning and Weaving Machines*

The roller spinning machine (invented in the 1770s) and the flyer-and-bobbin system developed for drawing wool to a more even thickness, dramatically increased productivity and the labor employment rate in the United Kingdom. In 1733, *John Kay* (1704–1779), a clockmaker, invented a simple weaving machine called the *Flying Shuttle*. He built it, supposedly, with nothing more than a pocketknife.

The Flying Shuttle improved on the old hand loom. A worker pulled a cord of rope back and forth to send a small piece of canoe-shaped wood, or shuttle, "flying" across a wood frame through threads to weave cloth. The machine only came into general use in the 1760s — after decades of trial-and-error — but once adopted, this first big invention in the textile industry doubled worker productivity. The invention was a small improvement and was still powered by people rather than coal, wind, or water. Nonetheless, it began the crucial process by which unskilled workers could produce more cloth with machines than skilled workers could produce by hand.

The next big challenge for industrial managers was to engineer a way that these new machines could be powered by an energy source that was more efficient and powerful than human muscle. In 1769, *Richard Arkwright* (1732–1792), a barber and wigmaker, figured out how to hook up a new spinning machine to a water wheel. Arkwright's spinning factory opened in 1771 along the River Derwent. It was an immediate success, spinning strong, high-quality threads cheaply and better than those spun by hand or a spinning jenny. Arkwright's cotton factory spun 24 hours a day, employing mostly women and children on 12 hour shifts. Each water frame spun 91 spools at a time, more than almost 100 people could spin on an old spinning wheel. Arkwright's business took off in a large part due to the assistance of the British Parliament, which outlawed the importation of superior cotton cloth from India to protect the new English textile industry.

Arkwright was not a great inventor. He borrowed most of his ideas from others. But he was one of the first and most successful entrepreneurs of the early Industrial Revolution; he understood the potential of these new textile inventions to produce inexpensive and high quality cloth. When others complained in court that he had stolen their ideas, Arkwright responded aggressively that "if any man has found a thing, and begun a thing, and does not go forwards ... another man has a right to take it up and get a patent for it." Though his attitude runs against the present patent law, he is still recognized as one of the most influential people of his time. He pushed the cotton industry to progress, spurring on the Industrial Revolution and creating great wealth for himself and for England. From 1800 to 1850, cotton products accounted for the majority of British exports (in monetary value).

1.5.3 *Steam Engines/Turbines*

Thomas Newcomen (1663–1729) invented a practical steam engine, and attempted to use it to pump water up from a coal mining tunnel in 1712. However, its fuel expenditure was significant due to its very low efficiency. *James Watt* (1736–1819) developed a new steam engine in 1782 (Fig. 1.9), which converted the pressure from the steam produced into a wheel rotation. The rotary type engine significantly expanded the areas for application for the steam engine;

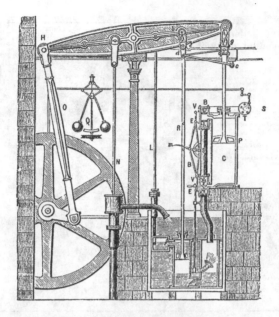

Fig. 1.9.   Engraving of a 1784 steam engine designed by Matthew Boulton and James Watt.

from the original water pump-up to general industrial power actuators [4]. Watt was the initiator of a modern machine with a *feedback control mechanism*. The use of the steam pressure indicator produced an informative plot of the pressure in the cylinder against its volume, which was helpful for maintaining a consistent rotation speed. Another important invention was parallel motion, which is essential in double-acting engines as it produces the straight line motion required to operate the cylinder rod and pump from the connected rocking beam, whose end moves in a circular arc. This was patented in 1784. These improvements, taken together, produced an engine which was up to five times as efficient in its use of fuel as the Newcomen engine.

Watt is known also as the initiator of the power unit (W, Watt) for power-driven machinery. He denoted power as "work per unit time," i.e., (force × velocity). The unit *horse power* (= 0.746 kW) was intentionally created to show how large a power his steam engine produced.

1.5.4 *Machine Tools*

The performance and machining accuracy of a machine is strongly dependent on the accuracy of each machinery part of the machine, and each part's accuracy cannot exceed the tool's cutting accuracy. Thus, the machine for making

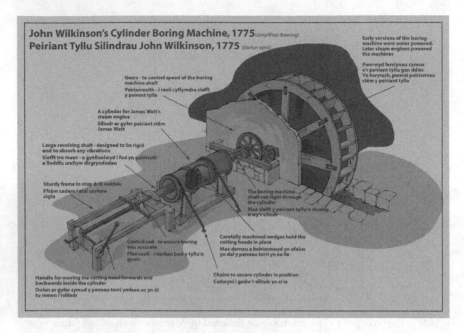

Fig. 1.10.   John Wilkinson's Cylinder Boring Machine 1775.

machines is called the "Mother Machine," which was highly respected during the Industrial Revolution. The manufacturing accuracy of cylinders and pistons are key in steam engines, where their precise motion is highly required. I would like to bring Wilkinson's cylinder boring machines and Maudslay's screw-cutting lathes, which contributed significantly to the production of steam engines, under the spotlight.

John "Iron-Mad" Wilkinson (1728–1808) was an English industrialist who pioneered the use and manufacture of cast iron and cast iron goods in the Industrial Revolution. He was the inventor of a *precision boring machine* (see Fig. 1.10) that could bore cast iron cylinders, such as those used in steam engines, which were used in James Watt's engines. Wilkinson's boring machine has been called the first machine tool. He also developed a blowing device for blast furnaces that allowed higher furnace temperatures, increasing their capacity.

Henry Maudslay (1771–1831) was a British machine tool innovator, who invented screw cutting lathes, and is considered a founding father of machine tool technology. One of his associates is *Joseph Whitworth* (1803–1887), who started the mass-production of machine tools such as boring machines and screw cutting lathes. In 1841, he devised the *British Standard Whitworth system*, which created an accepted standard for screw threads. Whitworth also created the

Whitworth rifle, often called the "sharpshooter" because of its accuracy. It is considered one of the earliest examples of a sniper rifle.

1.5.5 *American Mass-Production System*

The United States entered the Industrial Revolution in the late 19th century. After the American Civil War (1861–1865), the United States transformed from an agrarian society into an industrial society. The origin of America's industrial superiority over European countries can be found in American mechanical apparatuses' *capability in mass production*. In Europe's traditional manufacturing culture, individual products were manufactured on a small scale by a small firm with high manufacturing skills, as exemplified by the *Apprentice system*. A master (*Meister*) craftsman was entitled to employ young people as an inexpensive form of labor in exchange for providing food, lodging, and formal training in the craft. The system of apprenticeship was first developed in the Middle Ages and came to be supervised by *Craft Guilds* and town governments.

In contrast, mass-production and large-scale company management (i.e., the *American System*) originated from the US, where huge frontier lands and flexible human relationships enabled its birth and spread. There were several key components to the mass-production system: (1) part interchangeability and compatibility, (2) consumer adaptability, (3) an engineering-driven education system, and (4) the management/administration methods of large-scale firms.

Lieutenant General *Jean Baptiste Vaquette de Gribeauval* (1715–1789), a French artillery officer and engineer, introduced the so-called *Gribeauval system*, which is a Weapon Standard introduced through royal order in 1765. This system revolutionized the French cannon with a new production system that allowed the production of lighter, more uniform guns without sacrificing range. These guns contributed to French military victories during the Napoleonic Wars; this system was "arguably the best artillery system in Europe at that time," [11] based on the systems of organization and uniformity in ordnance.

The manufacturing technology for the rifled musket developed in France was transferred to the US because of the security alliance relationship between the two countries. However, the American Army initially could not duplicate the rifle, owing to their factories' inadequate cutting accuracy. America needed to wait until 1842, a half century after the French products, to self-mass produce the muskets by overcoming various technological barriers such as the standardization of design drawings, uniform materials, precise measuring scales, and administrative organization and/or management systems, including training programs for laborers, in order to produce compatible mechanical parts.

The United States had societal/cultural privileges that allowed America to overcome these problems, such as:

(1) A strong and wide consumer demand for a number of machine tools, because a small number of people needed to develop huge lands.
(2) Many retired army engineering officers started their spin-off firms, and manufactured machines (cutting machines, lathes) in the East Coast area, by adopting various standardized interchangeable parts.
(3) The development of administration and management methods, based on the Army's commanding manual on *Operations* (which taught on Army planning and orders production) and *Mission Command* (the protocol by which command and control of Army forces was achieved).

It is interesting to provide a brief history of my university, Pennsylvania State University, here. An originally agriculture-centric institution of higher learning, the university 'grew up' (i.e., *shifted its focus from agriculture to engineering*) following the larger transformative trend affecting the country. Penn State was founded in 1855 by Act P.L. 46, No. 50 of the General Assembly of the Commonwealth of Pennsylvania as the Farmers' High School of Pennsylvania. Centre County (located at the geological center of Pennsylvania) became the home of the new school when James Irvin of Bellefonte donated 809,000 m² of land. In 1862, the school's name was changed to the Agricultural College of Pennsylvania, and with the passage of the Morrill Land-Grant Act, Pennsylvania selected the school in 1863 to be the state's sole land grant college. In the following years, enrollment fell as the school tried to balance purely agricultural studies with a more classic education, falling to 64 undergraduates in 1875. Thus, the school's name changed once again to Pennsylvania State College. *George W. Atherton* became president of the school in 1882, and began working to broaden the school's curriculum. In 1886, the board of trustees approved the creation of a department of mechanical engineering. Shortly after, Penn State became one of the ten largest engineering schools in the nation.

1.6 Politico-Engineering

The intention of increasing national wealth and military strength increased international friction and led to the First and Second World Wars in the 20th century. Engineering of this period centered mainly around government-led production of weapons of war — this was the beginning of *Politico-Engineering*.

1.6.1 *École Polytechnique*

École Polytechnique was established in 1794 by mathematician *Gaspard Monge* during the French Revolution, and became a military academy under Napoleon I in 1804. Today, the institution still runs under the supervision of the French Ministry of Defense. Initially located in the Latin Quarter of Central Paris, École Polytechnique was moved in 1976 to Palaiseau on the Saclay Plateau, Southwest of Paris.

The French Revolution started on July 14, 1789 with the famous revolutionary motto: freedom, equality, and brotherhood, which are represented by the colors, blue, white and red, respectively, on their flag. After abolishing the monarchy in 1793, the revolutionary government pushed for idealistic public educational systems with a strong bias, such as the *absence of the need for scientists for the Revolution*. The new government consequently faced a serious lack of engineers, because most of the Royal and Combat Engineers of the time had defected from France to neighboring countries. The government decided to set up a new school to develop engineers urgently; hence the establishment of École Polytechnique.

Unlike conventional schools, which were separately established as institutes specializing in Civil Engineering, Mining Engineering, Naval Architecture, and Combat Engineering, École was designed to be a *polytechnique* university, i.e., one that taught engineering based on common basic sciences such as mathematics and physics. In fact, modern engineering began in École Polytechnique.

The original administrative ministry overseeing École Polytechnique was the Department of Interior. This changed after Napoleon's rise to power. The army, under the Corsican General *Napoleon Bonaparte* (1769–1821), became much more powerful. Napoleon set up a new government called the Consulate, with himself in power. This led to him becoming the dictator and, in 1804, the Emperor of France. Accordingly, the administrative ministry was switched to the Department of Defense; École Polytechnique became more directly controlled by the military.

The initial founding professors include *Gaspard Monge* (mathematician; 1746–1818), *Charles Augustin Coulomb* (combat engineer; 1736–1806), *Pierre-Simon Laplace* (mathematician; 1749–1827), *Joseph-Louis Lagrange* (mathematician; 1736–1813), *Jean-Baptiste Joseph Fourier* (mathematician; 1768–1830), and *Lazare Carnot* (combat engineer; 1753–1823). Their unique and radical student recruitment policy should be pointed out here. They copied the *Chinese Keju system*, a civil service Imperial Examination System in ancient times, through which officials were examined and selected [4]. It was first adopted in the Sui Dynasty (581–618) and lasted through the Qing Dynasty (1644–1911).

Any person could apply for the proficiency entrance examination conducted by École Polytechnique. Once accepted for entrance, the students were treated as government employees and tuition was paid for by the government. After Napoleon's reformation in 1804, École Polytechnique inclined more toward a military and elite-oriented character. The education curriculum shifted more toward analysis, and the total education period was reduced to two years. The École became known for the creation of extreme elites through a highly intensive educational curriculum — a system which was actually the most efficient "military education" system in the 19th and 20th centuries for industrialization engineering.

1.6.2 *The US Naval Research Laboratory*

Let me give a brief summary of the historical background behind how USA's military power was enhanced, from the following three viewpoints: (1) military human resources/Reserve Officers' Training Corps (ROTC), (2) education systems, and (3) advanced technology developments.

1.6.2.1 *Military Human Resource ROTC*

Because US territory (including the country boarder) has never been attacked by other countries — with the exception of the Japanese attack on Pearl Harbor during World War II in 1941 and al-Qaeda's attack on the World Trade Center in 2001 — unlike most European countries, the original role of America's military forces was close to that of the police, whose responsibilities were primarily to control the population drift from the "East" to the "West" after America's Independence. The largest war in US's history (in the 19th century), *The Revolution*, did not even attract the European military community, because it was just a domestic "civil war." The importance of US military power was only recognized during World War I, which started in 1914. However, at that time, the European countries had only expected the participation of the American economy and their military product supplies, and not the military itself. Though some military combat coops were sent to help British and French forces, they were directed by French commanders.

Based on the miserable experiences of World War I, the US recognized the importance of military officers with high combat knowledges and skills, who were able to effectively command soldiers. For this purpose, the Reserve Officers' Training Corps (ROTC) was formed, expanding the human resource supply chain to include "Reserve" and "Promoted uncommissioned/petty officers" — a big change from the original "military school graduates" only policy.

1.6.2.2 *Education Systems*

The United States has been maintaining the principle of *civilian control of the military* since the independence of the country. The President is the *Commander in Chief* of all military forces. The military budget, chief commanding officers, corps arrangement, and selection of military equipment are all determined by civilians, and these important management items are disclosed to Americans via politicians — with the exception of top secret items. Accordingly, there are many civilian military commentators who discuss military problems publicly through newspapers, journals, and TV programs. At the same time, the military cooperates with civilian reporters by granting interviews in order to transfer correct military information, so as to garner support from American citizens. This "civilian control" of the military stimulated *spontaneous civilian cooperation with military technology developments*, which was urgently required. The nuclear bomb created from the *Manhattan Project* initiated by *Albert Einstein* (Professor at Princeton University at the time), in collaboration with Los Alamos National Laboratory in 1945 is the best example of this. Penn State University is recognized as a Partner University to the US Navy, while Virginia Tech is a Partner University to the Army, because they have *Applied Research Laboratories* that are primarily sponsored by the Office of Naval Research and Army Research Laboratory, respectively. Because of the historical legacy of the "civilian controlled military," most US university professors are willing to work on military technologies, and receive research contracts from the various agencies of the Department of Defense.

1.6.2.3 *Advanced Technology Developments*

Americans were deeply worried about the great European war (World War I). When the Archduke of Austria-Hungary was killed in cold blood, igniting the most destructive war in human history, the initial reaction in the United States was the expected will for neutrality. As a nation of immigrants, the United States would have difficulty picking a side. Despite the obvious ties to Britain based on history and language, there were many United States citizens who claimed Germany and Austria-Hungary as their parent lands. Support of either the Allies (Britain, France, Russia, etc.) or the Central Powers (Germany, Austria-Hungary, etc.) might prove divisive. After 2.5 years of isolationism, the United States entered the Great War (WW I).

Thomas Alva Edison (1847–1931), who was felt that US lacked warfare technologies, argued that America should look toward science, when asked by a *New York Times* correspondent to comment on America's involvement in the War.

He proposed, in an interview published in May 1915, that "The Government should maintain a great research laboratory ... In [which] could be developed ... all the technique of military and naval progression without any vast expense." Secretary of the Navy *Josephus Daniels* seized the opportunity created by Edison's public comments to enlist Edison's support. Edison agreed to serve as the head of a new body of civilian experts — the Naval Consulting Board — to advise the Navy on science and technology. The Board's most ambitious plan was the creation of a modern research facility for the Navy. Congress allocated $1.5 million for the institution in 1916, but wartime delay and disagreements within the Naval Consulting Board postponed its construction until 1920.

The Laboratory's two original divisions, Radio and Sound, pioneered in the fields of high-frequency radio and underwater sound propagation. They produced communications equipment, direction-finding devices, sonar sets, and, perhaps most significant of all, the first practical *radar equipment* built in America. They also performed basic research, participating, for example, in the discovery and early exploration of the ionosphere (i.e., a region of Earth's upper atmosphere from an altitude of about 60–1,000 km, which is ionized by solar radiation). Moreover, the Laboratory was able to gradually work toward its goal of becoming a broadly-based research facility. By the beginning of World War II, five new divisions had been added: Physical Optics, Chemistry, Metallurgy, Mechanics and Electricity, and Internal Communications.

1.6.3 Technology Developments during World War II

As *Thomas Edison* predicted, science and technology (S&T) and engineering played a more significant role in determining the outcome of World War II than ever. Though many technologies were developed during the inter-war period of the 1920s and 1930s, several new key ones were developed in response to valuable lessons learned during the war, such as weaponry (e.g., weapons of mass destruction (WMDs)), logistic supports (surface, underwater, aerial vehicles), communications and intelligence, and medical technologies (surgical, chemical technologies), which might have ended the war a little earlier.

1.6.3.1 Weapon Developments

Weaponry includes new types of ships, land vehicles, aircraft, improved artillery, rocketry, and small arms. The Mitsubishi A6M Zero Fighter shown in Fig. 1.11 was developed by the Japanese Navy in 1940 as a new model of carrier-based aircrafts. The design concept was to reduce the weight of its parts as much as possible so as to realize maximum maneuverability under minimum engine power.

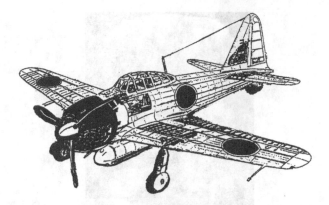

Fig. 1.11.   Japanese Mitsubishi A6M Zero Fighter (1940).

A fictionalized biography of *Jiro Horikoshi* (1903–1982) — the designer of the World War II fighter plane, the Mitsubishi A5M and A6M Zero — titled "The Wind Rises" was filmed by Studio Ghibli in 2013 (Director: Hayao Miyazaki); the reader may obtain some information on the planes' development history from the film. However, unlike the tale told in the film, serious disintegration accidents in midair occurred successively during the test flights. A young Marine engineer, *Tadashi Matsudaira* (1910–2000), was ordered to solve the problem. He discovered the cause of the accidents to be the abnormal twist flutter vibration on the wings, and solved the problem completely by modifying the wing design so as to strengthen their twist rigidity. These were the fighter planes that attacked Pearl Harbor, Hawaii, on the morning of December 7, 1941. They became the major threat against the US Air Force during World War II. Matsudaira adopted his own approach to suppressing self-excited vibration 20 years later (after the development of A6M Zero) to improve the Japanese *Shinkansen's* (i.e., bullet train) design [4].

Weapons of Mass Destruction (WMDs) include biological, chemical, and atomic weapons. Figure 1.12 shows the atomic bomb's mushroom cloud over Nagasaki in 1945. As many may know, the other Japanese city that had also been hit by an atomic bomb then was Hiroshima. I grew up in Hiroshima several years after the bomb. My memories of the period include my violin teacher having a serious face burn and missing one ear. I also remember visiting the Hiroshima Atomic Bomb Hospital many times to meet with my friends. This was perhaps the first war where military operations were aimed at the research efforts of the enemy. The exfiltration of *Niels Henrik David Bohr* (1885–1962) from German-occupied Denmark to Britain in 1943, the sabotage of Norwegian heavy water production, and the bombing of Peenemunde, were directly related with decelerating the German development of the atomic bomb.

Fig. 1.12.   Atomic bomb mushroom cloud over Nagasaki in 1945.

1.6.3.2 *Logistic Support Developments*

Logistical support created the vehicles necessary for transporting soldiers and supplies. These included trains, trucks, ships, and aircraft. The areas of development covered by *communications and intelligence* included devices used for navigation, communication, remote sensing, and espionage. Military operations were also conducted to obtain intelligence on the enemy's technology: for example, the Bruneval raid in 1942, for the German radar, and Operation Most III in 1944 (which obtained crucial intelligence on the German V-2 rocket). The surgical innovations, chemical medicines, *medical techniques*, and the technologies employed at factories and production/distribution centers in the area of industry during this era were also noteworthy.

1.6.3.3 *Military Technology Developments in Japan*

In contrast to the US' principle of *civilian control of the military*, in Japan before and during World War II, military information was not disclosed to civilians, but kept confidential and strictly regulated by the *Security Secret Information Regulation*. Under the name of Emperor Hirohito, the military uniform took the position of prime minister, and regulated discussions among the Japanese people on defense related issues. Accordingly, military, defense, and security issues unfortunately became a sort of taboo in Japanese society. Though the reasons for this have changed, even after World War II, the Japanese people still keep this "taboo" and do not want to discuss security issues.

After WWII, *General Douglas MacArthur*, on behalf of the United States, disassembled the Japanese military system in the *Japanese Constitution* to transform it into the Japan *Self-Defense Forces*. That means, officially, Japan does not have a military corps; only forces for self-defense. Second, because of the harsh regulations guarding the discussion of military strategy in public during World War II by the "military-controlled" government, the Japanese people became antagonistic toward the military, defense and even other crisis-related technologies. This degraded the power of the military even further 70 years after World War II — even though science and technology have improved Japan's economic progress. This was worsened by the Ministry of Education, Culture, Science and Technology (MEXT)'s discouragement of university professors from working on military, defense, and crisis technologies. It was a problematic attitude, which enhanced the damages during the Big Earthquake and Fukushima Nuclear Power Plant accident in 2011. For a detailed discussion on cultural ethics, see Chapter 7, Section 7.2, "Japanese R&D Ethics."

1.6.4 *Piezoelectric Application History*

The piezoelectric device is one of the best examples of technology developed by politico-engineering [5]. After the discovery of the direct piezoelectric effect (electrical charge/voltage generation under pressure) in single crystal quartz by the Curie brothers (*Pierre* and *Jacques Curie*) in 1880, no application development was made for more than 30 years.

It was at 11:45 pm on April 10, 1912 that the shipwrecking of the *Titanic* happened. As the reader knows well, this was caused by an iceberg hidden in the sea. If the ultrasonic sonar system had been developed, the incident could have been prevented. Though this tragic incident motivated (i.e., social demand) the development of ultrasonic technology using piezoelectricity, it did not generate enough investment for development.

World War I (immediately after the *Titanic* accident) created real investment for the acceleration of the development of ultrasonic technology, to enable the allied militaries to search for and detect *German U-Boats* under the sea (see Fig. 1.13). This was the strongest force from both social and political demands. *Paul Langevin* (1872–1946) a professor at the Collège de France in Paris (who had a wide circle of friends that included Drs. *Albert Einstein*, *Pierre Curie*, and *Ernest Rutherford*), in collaboration with the French Navy, started experimenting on ultrasonic signal transmissions into the sea. Langevin succeeded in transmitting ultrasonic pulses into the sea south of France in 1917. We can learn most of

Fig. 1.13.   German Navy U-Boat during WWI.

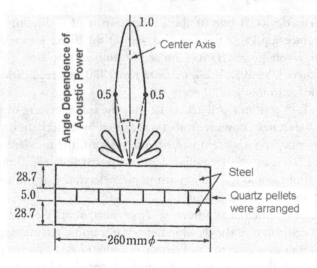

Fig. 1.14.   Original design of the Langevin underwater transducer and its acoustic power directivity.

the practical development approaches from this original transducer design (Fig. 1.14). First, 40 kHz was chosen for the sound wave frequency.

Increasing the frequency (shorter wavelength) leads to a better monitoring resolution of the objective; however, it also leads to a rapid decrease in reachable distance. Notice that quartz and Rochelle salt single crystals were the only available piezoelectric materials in the early 20th century. Since the sound velocity in quartz is about 5 km/s, 40 kHz corresponds to the wavelength of 12.5 cm in quartz. If we use a mechanical resonance in the piezoelectric material, a 12.5/2 = 6.25 cm thick quartz single crystal piece is required. However, in that period, it was not possible to produce such large high-quality single crystals [6]. In order to overcome this dilemma, Langevin invented a new transducer construction: small quartz crystals arranged in a mosaic were sandwiched by two steel plates. This sandwich structure is called the *Langevin type*, which is popularly utilized even nowadays (Fig. 1.14).

The next breakthrough in piezoelectric devices occurred during World War II. *Barium titanate (BaTiO₃, BT) ceramics* were discovered independently by three countries, United States, Japan, and Russia, during World War II; by *E. Wainer* and *N. Salomon* in 1942, *T. Ogawa* in 1944, and *B. M. Vul* in 1944, respectively. Compact radar system development required compact, high capacitance "condensers" (we used the term "condenser," rather than "capacitor" at that time). Based on the widely used "Tita-Con (Titania condenser)" composed of TiO_2–MgO, researchers doped various oxides to find higher permittivity materials. According to the memorial article authored by Ogawa and Waku [7], they investigated three dopants, CaO, SrO, and BaO in a wide fraction range. They found a maximum permittivity around the compositions of $CaTiO_3$, $SrTiO_3$, and $BaTiO_3$ (all were identified as a *perovskite structure*). In particular, the permittivity higher than 1,000 in $BaTiO_3$ was enormous (10 times higher than that in Tita-Con) at that time, as illustrated in Fig. 1.15. By the way, it is

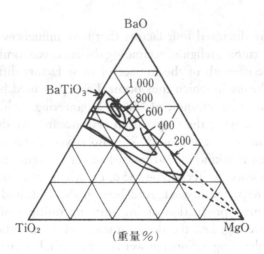

Fig. 1.15. Permittivity contour map on the MgO–TiO2–BaO system, and the patent coverage composition range (dashed line). Original Japanese article is used intentionally.

worthy to note that the first multilayer capacitor was invented with BT by Sandia Research Laboratory engineers with the coating/pasting method of BT slurry used in the *Manhattan Project* for the switch of the Hiroshima nuclear bomb [Private Communication with *Kikuo Wakino*, Murata Mfg].

It should be pointed out that the original discovery of $BaTiO_3$ had nothing to do with piezoelectric properties. Equally important are the independent discoveries by *R. B. Gray* at Erie Resister (patent applied in 1946) [8] and *Shepard Roberts* at MIT (published in 1947) [9]; that the electrically poled BT exhibited "piezoelectricity" owing to the domain re-alignment. At that time, researchers were arguing that the randomly oriented "polycrystalline" sample should not exhibit "piezoelectricity," but the secondary effect, "electrostriction." Because *R. B. Gray* is the first to verify that the polycrystalline BT exhibited piezoelectricity once it was electrically poled, he is deemed the "Father of Piezoceramics". The easiness in composition selection and in manufacturability of BT ceramics prompted *W. P. Mason* and others to study transducer applications using these electroceramics.

Following the methodology taken for the BT discovery, the perovskite isomorphic oxides, such as $PbTiO_3$, $PbZrO_3$, and $SrTiO_3$, and their solid solutions were intensively studied. The determination of the *lead zirconate titanate* $(Pb(Zr, Ti)O_3$ or PZT$)$ system phase diagram by the Japanese group, *E. Sawaguchi, G. Shirane,* and *Y. Takagi* [10], which led to the present PZT period, is particularly noteworthy. We will discuss further application developments using PZT in the following chapters.

1.7 Summary

In this chapter, we discussed four factors that have influences on engineering: STEP (i.e., social/culture/religion, technology/science, economics, and politics/law) and how the strength of the impact of these factors differ according to the situation of the era in which they occur. I also presented how the alchemy of the 16th century is an example of "Socio-Engineering": where religion controlled science by denying the Copernican (Heliocentric model) of our social system, while alchemy was approved. We also talked about how in the 17th and 18th centuries, mankind moved toward an era of engineering based on the motivations of Science & Technology (S&T); where (the so-called) Techno-Engineering was respected instead. And how in the 18th and 19th centuries, technologies for mass production at low manufacturing cost were required, and Econo-Engineering, i.e., the development and or invention of technologies essential to enhancing national power and strength, became the mainstream (see: *Industrial Revolution*). I also presented how the intention of increasing national wealth and military strength increased international friction, and eventually led to World Wars I and II in the 20th century, and how the engineering of this last period mainly comprised government-led production of warfare weapons — which was, in other words, the beginning of "Politico-Engineering."

1.8 Problem

The following popular technologies were originally developed by military institutes in the US and UK. Explore the historical background of each technology's development:

(1) World Wide Web (WWW).
(2) Global Positioning System (GPS).
(3) Digital Computer.

References

[1] G. Holton, *Introduction to Concepts and Theories in Physical Science*, Addison-Wesley Publishing, MA (1952).

[2] A. B. Arons, *Development of Concepts of Physics*, Addison-Wesley Publishing, MA (1965).

[3] S. Kisaka, *Inside Stories in Science & Technology History*, Nikkan Kogyo Newspaper, Tokyo (1986) [in Japanese].

[4] S. Miwa, *History of Engineering*, Chikuma Shobo Publ. Com., Tokyo (2012) [in Japanese].

[5] K. Uchino, The development of piezoelectric materials and the new perspective, Chapter 1 in *Advanced Piezoelectric Materials*, K. Uchino (ed.), Woodhead Publishing, Cambridge, UK (2010).

[6] K. Honda, *Ultrasonic World*, NHK books No. 710, Tokyo (1994) [in Japanese].

[7] T. Ogawa and S. Waku, *Splendid Tita-Bari*, Maruzen, Tokyo (1990) [in Japanese].

[8] B. Jaffe, W. Cook, and H. Jaffe, *Piezoelectric Ceramics*, Academic Press, London (1971).

[9] S. Roberts, Dielectric and piezoelectric properties of barium titanate, *Phys. Rev.* **71**, 890–895 (1947).

[10] E. Sawaguchi, Ferroelectricity versus anti-ferroelectricity in the solid solutions of $PbZrO3$ and $PbTiO3$, *J. Phys. Soc. Japan* **8**, 615–629 (1953).

[11] R. Chartrand, *Napoleon's Guns 1792–1815*, Osprey Publishing, Oxford (2003).

CHAPTER 2

Japanese and Global Technology Trend Change

This chapter reviews the change in trend for Japanese technology after World War II from the government/political view point. This analysis of the Japanese model is a good study for decision makers considering the politically-initiated engineering strategy — even for large countries like the United States and China.

After World War II, the recovery of basic infrastructures in Japan (that had been reduced to ruins by the war) was the most pressing domestic issue in Japan in the 1960s. After the 1980s, mass production technologies were sought after to produce consumer appliances in order to strengthen the country's power, resulting in the revival of a sort of "Econo-Engineering" under the direction of the political leaders until the 1990s. This raised Japan's economic status to No. 1 in the world in terms of GDP per capita (gross domestic product per person). However, the country's progress began to plateau in the 21st century.

2.1 Country Power Growth

The model of the country power change with year for Japan is visualized by using typical growth "S" curves in Fig. 2.1. The curves representing USA and China are also inserted with a time shift, taking into account the difference in the period of time in which they occurred; i.e., 20 years ahead of and 30 years behind Japan, respectively. Discussions in the following sections will be based on this illustration.

How can we measure a country's power? There are several factors to be considered:

- Economics
- Population/workforce
- Politics/global righteousness
- Education system
- Warfare technology
- Religion, etc.

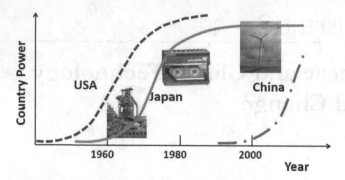

Fig. 2.1. Country power (GDP) change over year for Japan, USA (20 years ahead), and China (30 years behind) visualized by typical growth curves.

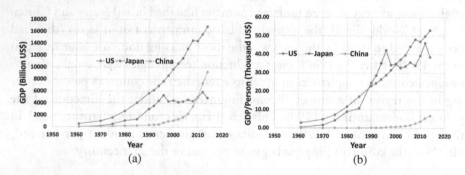

Fig. 2.2. (a) GDP and (b) GDP per capita change with year, for the USA, Japan, and China [1].

Figure 2.2 shows the (a) *Gross Domestic Product* (*GDP*) and (b) GDP per capita change relative to year for USA, Japan, and China. GDP may be considered one of the key factors for estimating a country's power [1]. These are the current top three (GDP) countries in the world.

2.1.1 *The USA*

The United States stays on top in both GDP and GDP per capita as of 2014. Though population growth rates in most developed countries are plateauing nowadays, the US is an exception to the rule, with its population still increasing constantly, almost linearly, year on year. This may be one of the origins of US's strength.

2.1.2 *Japan*

Japan's GDP drastically increased after the 1970s, though it was comparable with China's in the 1960s. The increase in GDP per capita at the end of the 1980s was significant, and it later exceeded the US's in the 1990s when Japan's economic strength was at its maximum in parallel to the strong Japanese Yen. At that time when the US desired to learn Japanese science and technology and the Japanese way of management that had raised the country's power miraculously, I was recruited by Penn State University, with a strong endorsement by the Department of Defense's (DOD) agencies. A big obstacle to Japan's continued success that arose after that was an issue pertaining to the population or work force. Japan's population growth decelerated gradually from 1980, and it actually started to decrease from 2010. The nation's GDP seemed to have reached saturation point during these several years.

2.1.3 *China*

China's GDP already exceeded Japan's in 2010, and may even exceed USA's in 2020, mainly owing to the inexpensive labor fee in China. However, China's GDP per capita is far behind the other two countries, even if we discount the intentionally controlled and devaluated Chinese Yuan. China has also exhibited a population growth deceleration after 2000. Population saturation is anticipated to be reached around the 2030s, which will be the bottleneck in Chinese development in the future.

The analysis above supports the country power growth curves illustrated in Fig. 2.1. We will now consider the detailed development tactics according to the initial, accelerated, and saturated stages in Japanese industries.

2.2 Initial Growth 1960s — Domestic Politics

2.2.1 *Heavier, Thicker, Longer, Larger*

Let me remind you of the famous revolutionary motto for the French Revolution; freedom, equality, and brotherhood, which are represented in their flag by the colors, blue, white, and red. A similar tactic was employed by the Japanese government for the product planning strategy in the development of Japanese industries. In the 1960s, the four-Chinese-character slogan used translates as "heavier, thicker, longer, and larger" (重, 厚, 長, 大); that manifested in the form of the manufacturing of heavier ships and thicker steel plates, and the constructing of longer buildings, and larger power plants (dams). These were Japan's key strategies for recovering from the ruins of World War II (i.e., *Domestic Politics*). Refer to the initial rise of the "S"-shaped growth curve in Fig. 2.1. As a symbol of Japanese recovery, the Tokyo Tower (the then highest tower in the world at 333 m) was built in 1958. This societal mood made me choose to join the electrical engineering department in the university to become a big water-dam engineer. However, the reader can understand why I changed this dream in the next section after entering the 1980s.

2.2.2 *Tokyo Summer Olympics*

A nationwide organization of an event (the Tokyo Summer Olympics) in October 1964 was planned by the government. Its political purpose was to advertise that the new Japan was no longer a wartime enemy, but a peaceful country. This transformation had been accomplished in 20 years. Though Japan's foreign policy was closely linked to the United States during the Cold War, Tokyo hosted the 1964 Summer Olympics in the spirit of peaceful engagement with the entire international community, including the Communist countries.

To host such a big event, Tokyo needed to modernize the infrastructure in time for the large number of tourists expected. Pushing the people's consensus in this direction, the government accelerated the plans to upgrade the city's physical infrastructure, including new buildings, highways, stadiums, hotels, airports, and trains. Multiple train and subway lines, a large highway building project, and the Japanese Bullet Train System (the Tokaido Shinkansen, the then fastest train in the world) were completed in 1964. Remember that the "Zero" fighter developers contributed significantly to this train development. Haneda International Airport and the Port of Tokyo were modernized. International satellite broadcasting was initiated, and Japan was connected to the world with a new undersea communication cable. Two symbols of this transformation, Tokyo Tower and Shinkansen Trains, are shown in Fig. 2.3.

2.2.3 *Industrial Pollution*

Unfortunately, the construction projects resulted in some environmental damage, the forced relocation of some residents, and the loss of some industries.

In addition, corruption by politicians and construction companies resulted in cost overruns and some shoddy work. Though I am aware of these political scandals, I will not discuss them because it is not relevant to S&T issues. Noting that the typical lifespan of cement is around 40 years, the reader can recognize that the infrastructures built in the early 1960s would have reached

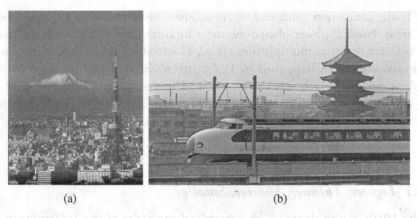

(a)　　　　　　　　　　　　　　　　(b)

Fig. 2.3.　Pictures of the two symbols of reformed Japan: (a) Tokyo Tower, built in 1958, and (b) a Shinkansen Train, established in 1964.

almost the limit of their durability, considering that over 50 years have passed since their construction in Japan. This is, at present, the most serious problem in any society concerned with sustainability. How can we keep or maintain the safety of our infrastructures? This is discussed in Chapter 6 (Sustainability Engineering).

Japanese society became wealthy because of the accelerated infrastructure development. However, subsidiary negative effects also started to surface: e.g., *industrial pollution.* The southern part of Japan, Minamata Bay, was contaminated by the disposal of a heavy metal (mercury) by Nippon Chisso Company (a chemical fertilizer manufacturer) — an act which resulted in the creation of thousands of *Minamata Disease* patients in that area. Sufferers of Minamata Disease suffered from neurological syndromes (e.g., insanity, paralysis, coma, and eventually, death).

Steel industries produced air pollution (e.g., yellow-tinted skies) and caused the rise of asthma among small children. As a famous proverb ("History repeats itself") predicts, China today is now facing the serious problem of a high PSI (Pollution Standard Index) concentration of fine particulate matter (PM 2.5), 40 years after the Japanese incident. The term *particulate matter (PM) 2.5* refers to tiny particles or droplets in the air that are 2.5 microns or less in diameter, generated by industrial activity. Particles in the PM 2.5 range are able to travel deeply into the respiratory tract and reach the lungs. Exposure to fine particles can cause negative short-term health effects such as eye, nose, throat, and lung irritation, coughing, sneezing, runny nose, and shortness of breath. Japanese industry researchers spent great efforts at reducing, trapping, and eliminating various PMs in the 1980s.

Traffic congestion generated severe acoustic noise pollution even in suburban areas. Nuclear power plants, the most high-tech industries, leaked hazardous radioactive wastes multiple times (e.g., The meltdown of Three Mile Island, Pennsylvania, which happened in 1979. Interestingly, I was involved in this incident during my first stay in Penn State University from 1978 till 1980 — this was my second nuclear-related experience after the bombing of Hiroshima. This experience is described in detail in Chapter 5 (Crisis Technology)).

2.3 Rapid Growth 1980s — Econo-Politics Revival

2.3.1 *Lighter, Thinner, Shorter, Smaller*

In the 1980s, a new slogan — the completely opposite of the 1960s slogan — began: "lighter, thinner, shorter, and smaller (軽, 薄, 短, 小)". Printers and cameras became lighter in weight, thinner computers and TVs (flat panel) gained popularity, printing time and information transfer periods became shorter, and air-conditioners and tape recorders (e.g., the SONY Walkman) were made smaller. Because of this societal mood, as a young university professor then, I started working on compact "piezoelectric actuators and motors" under my newly created terminology of "micromechatronics." Table 2.1 summarizes the technology development paradigm shift in Japan after World War II.

In order to enhance Japan's economic power, mass production technologies for consumer products were highly promoted, and "Econo-Engineering" continued until the end of the 20th century.

Table 2.1.   Technology development paradigm shift in Japan after World War II.

1960s	Heavier	Ship manufacturing
	Thicker	Steel industry
	Longer	Building construction
	Larger	Power plant (dam)
1980s	Lighter	Printer, Camera
	Thinner	TV, Computer
	Shorter	Printing time, communication period
	Smaller	Walkman, air-conditioner
2000s	Cooperation	Global standard — internet
	Protection	Defense technologies
	Reduction	Pb, Dioxin, CO_2, Energy consumption
	Continuation	Food supply, population–med technology

2.3.2 *Piezoelectric Devices*

Why did I focus on "piezoelectric actuators"? I created this terminology in the early 1980s, emphasizing on the interdisciplinarity link between electrical and mechanical engineering. The advantages of piezoelectric devices over electromagnetic (EM) types are summarized below [2]:

(a) More suitable for miniaturization

From the market research done by my team, tiny motors in the range of 5–8 mm are highly required for automated office and factory equipment. However, conventional EM motors with sufficient energy efficiency were rather difficult to produce, owing to significant Joule losses by the thin coil wires. Since the stored energy density of the piezo-device is larger than that of an EM type, micromotors 1/10 smaller in volume and weight can be achieved.

(b) No electromagnetic noise generation

Since magnetic shielding is not necessary, we can maintain a compact design.

(c) Higher efficiency

Figure 2.4 shows the efficiency vs. power relation for electromagnetic and piezoelectric devices (compiled from 3,000 data collected from commercial EM motor catalogs in 2000). The significant decrease in the efficiency of EM motors is mainly due to the Joule heat increase with each reduction of the coil wire's thickness. More than 95% of the input electrical energy in a wrist watch motor is spent generating heat! Since the size of the piezo-device has no effect on its efficiency, it is effective in the power range lower than 30 W.

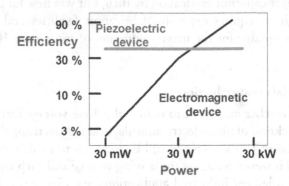

Fig. 2.4. Efficiency vs. power relation for electromagnetic (EM) and piezoelectric devices.

(d) Non-flammable

Moreover, piezo-devices are safer for device components dealing with electrical overloading or short-circuiting at the output terminal.

The first mass-production industrial application of piezoelectric actuators was a dot-matrix printer by NEC Corporation, Japan, which was birthed from my *invention of the co-fired multilayer actuator (MLA)* at Penn State University [3]. In the following couple of years, camera applications such as bimorph shutters by Minolta [4] and an autofocus mechanism with an ultrasonic motor by Canon [5] were widely commercialized. In parallel, piezoelectric linear actuators were being utilized for precise x–y stages in the late 1980s owing to the demand by semiconductor manufacturers [6]. Automobile applications of MLAs started in the 1990s; Toyota introduced the MLA to the damper, which is an electronic modulated suspension [7]. Siemens succeeded in using MLAs even in considerably severe high-temperature conditions for diesel injection control valves [8].

2.3.3 *Cost Minimization Strategy*

The piezoelectric ceramic actuator required a driving electric field of about 1 kV/ mm; i.e., a 1-cm thick sample needs 10,000 V directly (a dangerously high voltage). In order to achieve a low driving voltage and the miniaturization and hybridization of the devices, ceramic multilayer (ML) structures were invented by my group using the tape casting technique in the late 1970s. By reducing each layer's thickness by 1/10, we can reduce the driving voltage by 1/10 (while generating the same electric field level), leading to a practical drive voltage of 100 V for a 100 μm layer ML structure. Technologically, layers thinner than 3 μm are possible nowadays. Figure 2.5 compares the conventional cut-and-bond method with the tape-casting method. The tape-casting method had already been widely used for multilayer capacitor fabrication by then, but was new for piezo-actuator devices. Though it requires expensive fabrication facilities and sophisticated techniques, it is suitable for the mass production of more than 100,000 pieces per month.

(a) Raw materials cost reduction

A university researcher may work to reduce the drive voltage further by reducing the layer thickness of piezoelectric multilayer actuators using their improved technology, because 100 V seems to still be too high to be driven by a car battery (12 V_{DC}). However, reducing the driving voltage and each layer's thickness is not recommended for industrial applications, e.g., large size ML actuators for diesel injection valve control applications in diesel automobiles. Figure 2.6

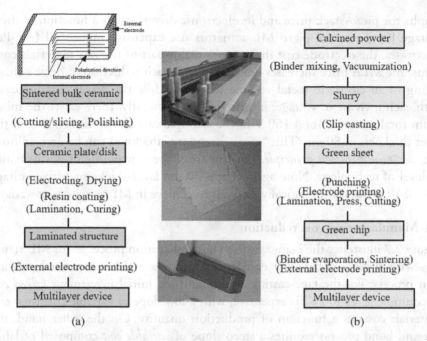

Fig. 2.5. Process comparison between (a) the conventional cut-and-bond method and (b) the tape-casting method.

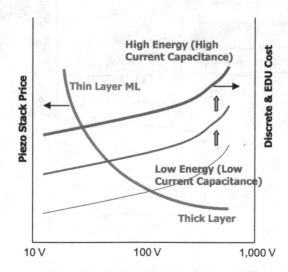

Fig. 2.6. Drive voltage dependence of piezo-stack price and the electronic driver cost.

graphs the piezo-stack price and its electronic driver cost as a function of drive voltage [9]. Since the piezo ML actuators use expensive rare-metal (Ag–Pd) electrodes, the electrode cost makes up a major part of the raw materials cost. Thus, the MLA price increases drastically with each reduction in drive voltage owing to an electrode metal volume increase, while the driver cost increases with each increase in voltage, in general. Consequently, there exists the minimum total cost of around 160 V in the example case shown in Fig. 2.6, at the layer thickness of 80 μm. This "sort-of-standard" thickness can be derived from the *cost minimization principle*, but not from the viewpoints of performance or level of technology. Note again that a 20 μm layer thickness (drive voltage 20–40 V) is not technologically difficult to achieve in ML actuators nowadays.

(b) Manufacturing cost reduction

Figure 2.7 illustrates the manufacturing cost calculation processes for ML actuators with automatic tape-cast equipment vs. a cut-and-bond manual production process. For the tape-casting equipment, the initial investment (*fixed cost* of equipment = $300 K) is expensive, with a low slope of variable cost (just raw materials cost) as a function of production quantity. On the other hand, the cut-and-bond process requires a steep slope of *variable cost* composed of labor fees and raw material costs. We can find an intersection between these two lines (point T in Fig. 2.7), which shows the product quantity that is the threshold

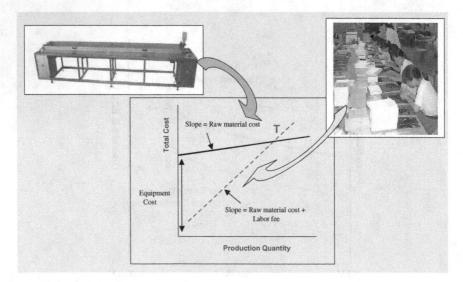

Fig. 2.7.   Total cost calculation comparison for the production of a multilayer product by automatic tape-cast equipment vs. manual cut-and-paste production.

above which the installation of automatic type-cast equipment starts to provide a better profit.

Because labor fees are expensive in developed countries, the introduction of a tape-casting facility is usually recommended when the production quantity exceeds 0.1 million pieces per year. However, due to the lower labor cost in developing countries such as Thailand and China, this production threshold quantity is dramatically high; at about 2 million pieces in multilayer (ML) actuator production. If the production quantity in these countries lies at 1 million pieces or less, the manual production process is actually the cheaper option [10]. This was the motivation behind Japanese manufacturers' *globalization* aspirations to move their factories to or find OEM partners in a foreign country (i.e., *off-shore production*) in the 1990s.

(c) Currency exchange rate — Yen

Another key to a country's economy to be considered is its *currency exchange rate* (see Fig. 2.8). As the reader may know, by the early 1960s, the US dollar's fixed value against gold under the *Bretton Woods System* of fixed exchange rates was seen as overvalued. A sizable increase in domestic spending on President Lyndon Johnson's Great Society programs and a rise in military spending caused by the Vietnam War (1965–1973) gradually worsened the overvaluation of the US dollar. The Bretton Woods system dissolved between 1968 and 1973. In August 1971, US President Richard Nixon announced the "temporary" suspension of

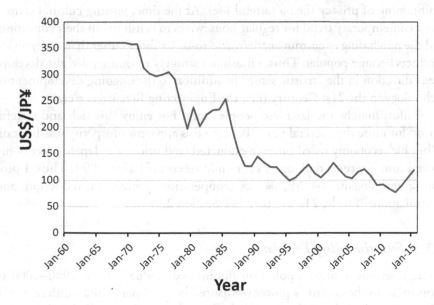

Fig. 2.8.   Change in Japanese Yen currency exchange rate against US$ year on year (1960–2015).

the dollar's convertibility into gold. An attempt to revive the fixed exchange rates failed, and major currencies began to float against each other.

The Japanese Yen (JPY) currency exchange rate was fixed against the US dollar (USD) as 1 US$ = 360 JP¥ until 1970. However, after the collapse of the Bretton Woods system, the Yen started to increase its value, producing what appeared to be a dramatic increase in GDP. During my time as an assistant professor (1975–1985), 1 US$ = 220–230 JP¥. Though it still felt slightly expensive to visit the US then, it was no longer just a dream anymore. Furthermore, this allowed us to enjoy purchasing various imported goods, e.g., alcohol, tobacco, chocolate, branded apparel, etc. [10].

2.3.4 *Beautiful, Amusing, Tasteful, Creative*

At the end of 1980s, when the Japanese economy caught up with the US, and Japan's GDP per capita exceeded the US's (partially due to another big jump of the Yen currency in 1985–1987 up to 1 US$ = 130 JP¥), Y. Hirashima indicated a new consumer attitude and new social trends in the 21st century in his book [11]. Compared to the technological trends in the 1980s aspiring to create lighter, thinner, shorter and smaller technologies, he proposed for "美, 遊, 潤, 創 (beautiful, amusing, tasteful, and creative)" to be the slogan for the 2000s [12]. Well-known brand apparel (such as Louis Vuitton) are really beautiful, while TV gameshows are merely amusing products. Mobile/cellular phones are no longer only communication tools anymore, but also a kind of accessory that allows the maintaining of privacy (i.e., a tasteful life). At the time, visiting cultural centers was a contemporary trend for regular housewives to brush up on their education, and the purchasing of custom-made/made-to-order shoes rather than ready-made products became popular. Thus, Hirashima strongly recommended the development direction in the "artistic sense" in addition to the ongoing development of technology in the 21st Century (i.e., the Engineering Renaissance era).

Unfortunately, the Japanese people could not enjoy this rich and tasteful period for more than several years. Because this apparent prosperity was built on a "bubble" economy based on over-evaluated land prices; the Japanese economy moved into a serious dark age ("economic recession") after 1990. Thus, I propose a new slogan, "協, 守, 減, 維 (cooperation, protection, reduction, and continuation)" for the 21st century (see Section 2.4).

2.3.5 *Sustainability Problems*

Though serious industrial pollution diminished gradually from 1980–2000 in proportion to the country's power (measured in GDP per capita), different subsidiary effects had already begun: (1) the "greenhouse effect" and global warming due to CO_2 emissions by over-produced automobiles, (2) the energy crisis due to the global over-consumption of energy, and limited fossil energy sources (oil) in addition to political mismanagement, and (3) population growth due to advanced medical technologies. Though longer life spans are typically welcomed by individuals (now the average life expectancy is approaching 87 years for the average Japanese female, and 81 for males — people of a nation already known for having the world's longest human life expectancy), the over-population of humans on earth will create an imbalance in nature (relative to other animals), and majority senior population in particular, causes societal and economic problems (e.g., issues with pension, health insurance, work force, etc.).

2.4 Growth Maturing 2000s — International Politics

When the 21st century began, as a consequent result of the rapid industrialization that happened before, environmental degradation, resource depletion, and food famine have become major problems. Global regimes/regulations have been strongly called for, and government-initiated technology, i.e., *politico-engineering*, has become even more important in order to satisfy these regulations. I would like to propose in this book a new four-Chinese-character slogan for the "era of politico-engineering": 協, 守, 減, 維 (*cooperation, protection, reduction,* and *continuation*) (Table 2.1). Global coordination and international cooperation in the standardization of internet systems and computer cables became essential to accelerating mutual communication.

The *Kyoto Protocol* (adopted in December 1997) is an international agreement linked to the United Nations Framework Convention on Climate Change [12]. Its major feature is to set binding targets for 37 industrialized countries and the European community to reduce greenhouse gas emissions. *Protection* of a country's territory and environment from enemies or natural disasters, and the spread of infectious diseases is mandatory. The *reduction* of the use of toxic materials such as lead, heavy metals, dioxin, and of the use of resources and energy consumption are also key; and the *continuation* of society, i.e., *status quo* or the establishment of a *Sustainable Society*, is important to promote.

Considering the current situation in the world, such as terrorist attacks, territorial aggression, major disasters (natural and man-made), we will discuss *crisis technologies* in addition to *sustainability technologies* as possible avenues of *politically-initiated engineering* in this book.

2.5 Summary

This chapter reviewed the Japanese technological trend change after World War II from the political view point of the government. We discussed how the

Japanese government employed slogans to direct the product planning strategy in Japanese industries. In the 1960s (initial growth period), the four-Chinese-character slogan was "heavier, thicker, longer, and larger (重, 厚, 長, 大)," i.e., manufacturing heavier ships, and thicker steel plates, and constructing longer buildings and larger power plants (dams) to recover from the ruins of World War II. A completely opposite slogan started in the 1980s during the rapid growth period, i.e., "lighter, thinner, shorter, and smaller (軽, 薄, 短, 小)." Printers and cameras became lighter in weight; thinner computers and flat panel TVs gained popularity; printing time and information transfer periods became shorter; and air-conditioners and tape recorders became smaller. We also covered how, despite improvements in technology, GDP and life expectancy, environmental degradation, resource depletion, and food famine became major problems when we entered the 21st century. I talked about how government-initiated technology, i.e., *politico-engineering*, has become important in order to satisfy/abide by new global regulations, and proposed a new four-Chinese-character slogan for the era of "politico-engineering," i.e., "協, 守, 減, 維 (cooperation, protection, reduction, and continuation)." Mention on the necessity for international *cooperation* in the standardization of internet systems and computer cables in order to accelerate mutual communication, was made; as well as how the Kyoto Protocol is an international agreement linked to the United Nations Framework for reducing greenhouse gas emissions; and how the *protection* of sovereign territories and environment from enemies or natural disasters, and the spread of infectious diseases is mandatory. I also argued that a *reduction* in the use of toxic materials such as lead, heavy metals, and dioxin, and of the use of resources and energy consumption are also key to society *continuation* either by maintaining the *status quo*, or the development of a *Sustainable Society*.

References

[1] http://www.tradingeconomics.com/country-list/gdp.

[2] K. Uchino, *Micromechatronics*, CRC/Dekker, NY (2003).

[3] K. Uchino, S. Nomura, L.E. Cross, R.E. Newnham and S.J. Jang, Electrostrictive effect in perovskites and its transducer applications, *J. Mater. Sci.* **16**(3), 569–578 (1981).

[4] Minolta Camera, Product Catalog "Mac Dual I, II" (1989).

[5] T. Sashida, *Mech. Automation Jpn.* **15**(2), 31 (1983).

[6] S. Moriyama, T. Harada and A. Takanashi, *Precision Mach.*, **50**, 718 (1984).

[7] Y. Yokoya, *Electronic Ceramics*, **22**(111), 55 (1991).

[8] C. Schuh, *Proc. New Actuator 2004* (Bremen, June14–16), 127 (2004).

[9] A. Fujii, *Proc. Japan Tech. Transfer Assoc.* Meeting on Dec. 2, Tokyo (2005).

[10] K. Uchino, *Entrepreneurship for Engineers*, CRC Press, NY (2010).

[11] Y. Hirashima, Product Planning in Feeling Consumer Era, Jitsumu-Kyoiku Publ., Tokyo (1986). [In Japanese].

[12] http://unfccc.int/kyoto_protocol/items/2830.php.

CHAPTER 3

Global Politics in Engineering

The importance of "Politico-engineering" is increasing in the present 21st century. "Politico-engineering" or "politically initiated engineering," (i.e., engineering initiated by various levels of government — municipal, state, country or union — according to the need and/or emergency, to overcome the problematic issue) is occasionally under strong constraints of international relationships, such as wars, global regimes, etc. This chapter considers the fundamentals of global regimes in terms of science and technology. The strategic decision-making principle is introduced based on the simplest *Game Theory*. Then, typical model regimes are described. More complex *foreign policy analysis* is discussed with advanced game theories in Chapter 8. This part is highly in debt to a Japanese book authored by Yoshinobu Yamamoto [1].

3.1 Levels of Systematization

We will consider here, levels of systematization of multiple regions, states or countries. Table 3.1 summarizes four types of system levels in terms of (1) the *norm/rule* for common problems, (2) *governance organization*, and (3) the *unification of security strategy/diplomacy* [1].

Anarchy in this section means that each state behaves independently without regime, but does not mean that they are in a conflict or at war. The base of the relationship is inter-governmentalism, i.e., each state promotes its own freedom or government system. There is no systematization in terms of the above points (1), (2), or (3).

The next level is a *global regime*, where all the states involved have a common problem, and they set mutual norms or rules to solve this problem, and create a shared regime for strategic cooperation. The *Asia-Pacific Economic Cooperation (APEC)* is an example of such a regime. APEC is a regional economic forum established in 1989 to leverage the growing interdependence of the Asia-Pacific. APEC's 21 members aim to create greater prosperity for the people of the region by promoting balanced, inclusive, sustainable, innovative, and secure growth, and by accelerating regional economic integration. APEC

Table 3.1.   Levels of regional systematization.

System Level	Element			
	Norm/Rule for a Common Problem	Governance Organization	Unification of Security Strategy/Diplomacy	Example
Anarchy	×	×	×	
Regime	○	×	×	APEC
Government	○	○	×	EU
Federation/Confederation	○	○	○	USA

operates as a cooperative, multilateral economic and trade forum. Members participate on the basis of open dialogue and respect for the views of all participants. In APEC, all members have an equal say and decision-making is reached by consensus. There are no binding commitments or treaty obligations. Commitments are undertaken on a voluntary basis and capacity building projects help members implement APEC initiatives. Though there is a permanent secretariat based in Singapore, no solid governance organization exists.

The third level is a *government*, where a solid organization (an entity that is more than a shell organization, with a budget, personnel, and secretariat) determines the group decision and executes a practical action for a certain problem. We may call such an organization a "government." This government focuses only on particular problems, excluding occasional military and diplomatic areas. The *European Union* (*EU*) may be an example of such a system. Though the EU initially targeted to become a federation/confederation like the United States of America, there still exists a big gap/conflict between the EU Council and its composing countries. The EU's unique institutional set up includes the following [2]:

(1) the EU's broad priorities are set by the *European Council*, which brings together national and EU-level leaders,
(2) directly elected Members of European Parliament represent European citizens in the *European Parliament*,
(3) the interests of the EU as a whole are promoted by the *European Commission*, whose members are appointed by national governments, and
(4) governments defend their own country's national interests in the Council of the EU.

There are three presidents in the EU: President of the European Parliament, President of the European Council, and President of the European Commission. There also exists the European Parliament, but their power and appearance fall far behind that of a federation, particularly in issues of high politics, such as security alliances and diplomacy.

The highest level of systematization is called a *federation/confederation*, where all the composing states take common strategies in all areas, including military and diplomatic issues. The best example of this is the United States of America.

In this book, we focus on the necessity for a global regime and the setting of procedure to create a kind of norm/rule for a common problem among multiple states/countries, so as to achieve global cooperation — particularly in the development of crisis and sustainability technologies.

3.2 Global Regime — Fundamentals

We consider how the regime is established, what type of regime is generated, and how the regime is maintained or modified among the stake-holders. We will use three key factors of the creation, maintenance, and modification of a regime, as illustrated in Fig. 3.1 [1], to evaluate the regime: power, profit, and norm.

The *power* of a state/country based on military and economic force (including market size), and controlling capability of the opponent state, may be a base on which to create a regime. This is exemplified by a regime set by the hegemony of allied countries, such as the North Atlantic Treaty Organization (NATO). NATO is an intergovernmental military alliance signed in 1949. The organization constitutes a system of collective defense whereby its member states agree to mutual defense in response to an attack by any external party (e.g., the then-Soviet Union (USSR) and its related communist countries, at the time of NATO's inauguration). Twenty-eight states across North America and Europe make up the members of NATO. I call this power concept *realism*.

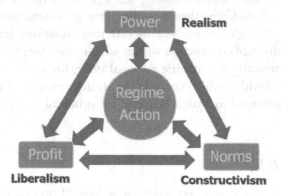

Fig. 3.1.   Three factors to the creation, maintenance, and modification of a regime: power, profit, and norm.

Profit is the second factor, where the profit creation mechanism is the origin of a regime. The *General Agreement on Tariffs and Trade* (GATT) is an example of such an arrangement. Preceding the *World Trade Organization* (WTO), GATT was a multilateral agreement regulating international trade. Its purpose was the "substantial reduction of tariffs and other trade barriers." It was negotiated during the United Nations Conference on Trade and Employment in 1947 and later replaced by WTO in 1995. This profit-oriented concept is called *liberalism*.

The third factor is not based on materials such as military or economic power (i.e., *hard power*), but mental, psychological or societal *norms* (i.e., *soft power*), such as justice or righteousness. This concept is sometimes called *constructivism*. The Treaty on the Non-Proliferation of Nuclear Weapons (*NPT*) is an example of such a multilateral treaty. Established in 1970, its goals are to prevent the spread of nuclear weapons and weapons technology; promote cooperation in the peaceful uses of nuclear energy; and further the goal of achieving nuclear disarmament.

The reader should realize that a regime and its consequent action are dependent on these three independent parameters (power, profit, and norm) in a complex function, which is not a simple linear relationship at all!

Most "hard power" is directly related to developments in Science and Technology (S&T), while "soft power" leans more toward political leaders' psychological norms and cultural milieu. However, like the NPT, the regimes treat many issues that originate from S&T developments since, for advancements such as nuclear technologies, it is difficult to distinguish between weaponized and peaceful applications.

3.3 Global Regime — Categorizations

We consider a *rational decision-making* strategy based on the simplest *Game Theory* in this section. Game theory studies strategic interaction between individuals mainly through the analysis of different situations (called "games"). A "game" is a fully explicit structure within which *actors* (*or players, agents*) conduct *actions* strategically to maximize personal *utility* (or *payoff*). Games provide a simplified world within which to study the strategy used and decision-making process, and is also used to analyze real and hypothetical international relations scenarios.

3.3.1 *Payoff Table*

Since the model of the *prisoners' dilemma* is best illustrated by the story for which it is named, we start by learning the basic idea behind this model from the story. Suppose that a serious crime is committed, and our offenders, Bonnie and Clyde, are apprehended and questioned separately by the police. Because the police do not have enough evidence to convict the pair on the principle charge, a confession is important. There are three possible scenarios:

(1) If neither confesses, the police can attempt a conviction on lesser charges (which could result in a two-year sentence for both parties).
(2) If one suspect confesses to the crime and testifies in court against the other, the snitch will only receive a one-year sentence, while the other will get 12 years.
(3) If both confess, each receive a six-year sentence.

Table 3.2 shows the four possibilities in a table called a *payoff table*. A payoff table shows how much the actor can make for a possible combination of actions. Table 3.2 illustrates the simplest case for two actors/players for two action items (thus, $2 \times 2 = 4$ possibilities). When Player 1 (Bonnie) chooses Action 1 ("Don't Confess"), there are two possibilities for Player 2 (Clyde); Action 1 ("Don't Confess") and Action 2 ("Confess"). Cell A shows a pair of payoffs (in this case, a "negative payoff" or penalty, i.e., the smaller the better) for Player 1 and Player 2: the "xxx" in "xxx/yyy" corresponds to the payoff for Player 1, and "yyy" corresponds to the payoff for Player 2. Similarly, when Player 1 (Bonnie) chooses Action 2 ("Confess"), there are two possibilities for Player 2 (Clyde); Action 1 ("Don't Confess") and Action 2 ("Confess"). Cell C shows a pair of payoffs for Player 1 and Player 2. The payoff for Bonnie is a one-year sentence, while Clyde receives the 12-year sentence. Let us now consider a *rational decision-making* process:

(1) If the two parties can come to an agreement/collude, both will choose "Don't Confess" (Cell A), leading to the outcome of a two-year sentence for both.
(2) If an agreement is not possible/allowed (for example, in a free market, collusion is domestically illegal), Bonnie will consider the following: (a) when Clyde takes "Don't confess," it is more advantageous for Bonnie to choose

Table 3.2. Payoff table for the prisoners' dilemma.

	Player 1: Clyde	
Player 2: Bonnie	Don't Confess	Confess
Don't confess	A: 2 Years/2 Years	B: 12 Years/1 Year
Confess	C: 1 Year/12 Years	D: 6 Years/6 Years

"Confess" (Cell C) because the sentence will be reduced from two years (Cell A) to one year (Cell C). (b) When Clyde takes "Confess," Bonnie had better take "Confess" (Cell D) because the sentence will be six years for both parties — shorter than 12 years (Cell B). In both (a) and (b) cases, it is more advantageous for Bonnie to choose "Confess."

(3) If an agreement is not possible, Clyde will consider exactly the same logical train of thought as long as Clyde is a rational person. In all the possible scenarios, it is in Clyde's best interest to choose "Confess".

Therefore, we can say that Cell D is the *dominant strategy* for both Bonnie and Clyde. This is called as the *Nash Equilibrium* solution. *John Forbes Nash, Jr.* (1928–2015) proposed this non-cooperative game theory in 1950. Note that the dominant strategy (a six-year/six-year sentence) is not the best solution for either players, in comparison with Cell A (two-year/two-year sentence).

It should also be noted that this theory is based on the assumption that all the players are rational, and that they engage in rational decision-making *to maximize personal utility* (or payoff). If Bonnie loves Clyde very much, she may consider reducing Clyde's penalty her highest priority, and sacrifice herself. In this scenario, Bonnie will choose Cell B ("Don't Confess") in order to minimize Clyde's penalty to only one year. Similarly, if Clyde loves Bonnie very much, he will choose Cell C ("Don't Confess"). Consequently, the final solution will be Cell A ("Don't Confess" and "Don't Confess"), i.e., a situation of a two-year sentence for both Bonnie and Clyde.

Supposing that the ideal solution should be Cell A for both players 1 and 2, we can obtain two strategies to reach this scenario:

(1) An agreement/collusion between players 1 and 2 to choose "Don't Confess." This corresponds to the setting of a "regime" between two states/countries. There may be two cases in which this arrangement is possible: one is a profit-based regime where two parties hold equal power, while the other is power-based regime where one player dominates over the other in terms of one country having stronger power. If Clyde holds power over Bonnie (such as in cases of domestic violence), Bonnie needs to obey Clyde's decision not to confess; a situation which may be analogous to the *hegemony* between allied countries. OPEC is a good example of such a profit-based regime, as introduced in the Prologue in Chapter 1.

(2) Set a norm-based naturally-generated agreement/regime, so that the two players can escape from the dominant strategy, i.e., the worse scenario (six-year sentences for both), shown in Cell D. Environment protection

regimes may correspond to this category, which is not directly related with the maximization of personal profit, but with maintaining the common environment of all players.

3.3.2 Non-Regime Type Games

We consider six typical 'games' in game theory between two players, Players A and B, with two action possibilities, "Action 1" and "Action 2" in the following two subsections. The games are: (a) harmony, (b) prisoner's dilemma, (c) coordination, (d) the stag hunt, (e) zero-sum, and (f) independence game. Table 3.3 summarizes the payoff tables for these six games. We discuss the three non-regime type games, (a) harmony, (f) independence, and (e) zero-sum games, first.

3.3.2.1 Harmony Game

Refer to Table 3.3(a). The best profit for both Players A and B are realized by a status in Cell A, i.e., "Action 1" and "Action 1." From the payoff viewpoints of

Table 3.3. Payoff tables for six typical games: (a) harmony, (b) prisoner's dilemma, (c) coordination, (d) stag hunt, (e) zero-sum, and (f) independence.

(a) Harmony

		B Action 1	B Action 2
A	Action 1	(4, 4)	3, 2
	Action 2	2, 3	1, 1

(b) Prisoners' Dilemma (Collaboration)

		B Action 1	B Action 2
A	Action 1	3, 3	1, 4
	Action 2	4, 1	(2, 2)

(c) Coordination

		B Action 1	B Action 2
A	Action 1	(3, 4)	1, 1
	Action 2	2, 2	(4, 3)

(d) Stag Hunt

		B Action 1	B Action 2
A	Action 1	(4, 4)	1, 3
	Action 2	3, 1	(2, 2)

(e) Zero-Sum

		B Action 1	B Action 2
A	Action 1	4, 1	(2, 3)
	Action 2	3, 2	1, 4

(f) Independence

		B Action 1	B Action 2
A	Action 1	3, 2	(3, 3)
	Action 2	2, 2	2, 3

◯ Nash Equilibrium

both players, we cannot find any motivation to adopt another combination for either player. In this case, there is no need for any negotiation or regime to realize the best solution; this is referred to as a state of *harmony*.

3.3.2.2 *Independence Game*

Refer to Table 3.3(f). The best profit for Player A is obtained when A takes "Action 1," regardless of whether Player B takes "Action 1" or "Action 2." While the best profit for Player B can be realized when B takes "Action 2," regardless of the action by Player A. In other words, the *dominant strategy* for Player A is Action 1, while for Player B, it is Action 2. The action of each player does not affect the other player's action, and each can independently decide on the action that maximizes his utility or profit, i.e., there is no mutual interaction. As a result, the most probable status will be Cell B, where Player A chooses "Action 1" and Player B, "Action 2." Both players will receive the maximum profit among all possible scenarios. Since neither player has the motivation to opt for other options apart from Cell B, they do not need to establish any regime to realize a better solution.

3.3.2.3 *Zero-Sum Game*

Regime setting may also be difficult in Table 3.3(e), where the two actors/players are completely antagonistic in terms of their profits (i.e., a *zero-sum* game). A combination of "Action 1"–"Action 1" (Cell A) provides the best profit to Player A, but the worst profit for B; while "Action 2"–"Action 2" (Cell D) realizes the best profit for Player B, but the worst profit for A. Note that the sum of the utilities of two players in all cells (or action pairs) is exactly the same (i.e., zero-sum). Thus, it is impossible to generate a regime/agreement between these two players to take a coherent action (1 or 2). Furthermore, the better profit for Player A is obtained when A takes "Action 1," regardless of the action taken by Player B. While, the better profit for Player B can be realized when B takes "Action 2," regardless of the action by Player A. As a result, the equilibrium solution will be in Cell B, where A takes "Action 1" and B, "Action 2" (similar to the result in an *Independence Game*). However, note that unlike the *Independence Game*, the profits in Cell B are not the best for either player A or B, leading to no good solution between these players under any agreement or regime. This game example can be found in cases of territorial invasion. Two adjacent countries, A and B, fight to obtain a border area. Action 1 means that the border area belongs to country A, and Action 2 results in the border area belonging to country B.

Even though Cell B is temporarily stable (this status may be represented by the sharing of the border area by countries A and B), each country will try to shift the status to the one most beneficial for themselves (i.e., Cell A for country A and Cell D for country B, respectively). Thus, the conflict will continue without reaching an agreement.

The above three cases do not generate a regime (they are non-regime types). We will now consider the remaining three cases that may generate a regime; in which the setting of a regime/rule will increase the profit in comparison with the non-regime scenarios, where each actor/player behaves independently, or even start fighting in a zero-sum game. In regime types, the sum of the utilities of all players in all cells (or action combinations) is equal to or more than the value in a stable "Nash Equilibrium" via the dominant strategy (i.e., *positive-sum*).

3.3.3 Regime Type Games

Figure 3.2 illustrates when a regime should be created on the power–profit–norm map. We explained three non-regime types above: (1) where the states/countries (i.e., players) are strongly antagonistic, (2) when the relation of the states are *independent*, and (3) when the states have a relationship with consistent interests (*harmony*), though these are rare in practice. A regime can be created under the following conditions: (1) when the concerned states are *dependent* each other and (2) their interests are *between antagonistic and harmonious*, i.e., mixed motivation between interest conflict and corporation.

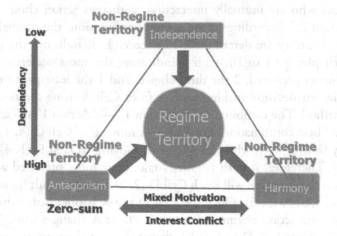

Fig. 3.2. Metaphorical illustration of "regime territory" among three factors to regime creation: power, profit, and norm.

During the Cold War, the US and the former Soviet Union (USSR) kept a rather antagonistic relation with no security treaty; while the *North Atlantic Treaty Organization* (*NATO*) and the *Warsaw Pact* (The Warsaw Treaty Organization) were established in the western and eastern territories respectively. NATO and the Warsaw Pact were inter-governmental political and military alliances that illustrate regimes formed under power-driven scenarios; NATO (established in 1949) involved the United States, Canada, and Western European nations, and the Warsaw Pack (established in 1955) comprised of the Soviet Union and several Eastern European countries.

Examples of profit-driven regimes (established in order to accelerate free trade) include multilateral trade agreements such as the then *GATT* (now WTO). These treaties aimed to abolish quotas and reduce tariff duties among contracting nations.

An example of a regime birthed in a hope to establish a "norm" (i.e., normal set of rules/procedures to deal with common problems) is the multilateral *NPT* treaty (i.e. Treaty on the Non-Proliferation of Nuclear Weapons; see Section 3.4.2.1 for more information) which entered into force in 1970 with 190 parties, including the five nuclear-weapon States (US, Russia, UK, France, and China).

3.3.3.1 *Prisoners' Dilemma* (*Collaboration Game*)

Though we have explained the *Prisoners' Dilemma* (*Collaboration Game)* in the former section, we reconsider this game here. Refer to Table 3.3(b). In it are two players/actors who are mutually interacting, with two action choices: "Action 1" and "Action 2." According to the action combination, the mutual relationship and each utility are determined. The score, 1–4, indicates the preference level of each player (A or B), with 4 indicating the most preferred action, 3 the second-most preferred, 2 the third-choice, and 1 the least-preferred choice. Let us begin our decision-making process from Cell A using a rational/logical thinking method. The combination of "Action 1" — "Action 1" in Cell A (3, 3) is the second best combination. For Player A, moving to Cell C (4, 1) increases their utility from 3 to 4; while for Player B, moving to Cell B (1, 4) increases their utility. Therefore, if both countries take their most preferred action, the final solution combination will reach Cell D (2, 2), a worse result for each player compared to Cell A (3, 3). Once the Cell D status is obtained, neither player will change their action anymore, because if Player A changes from "Action 2" to "Action 1" (from Cell D to Cell B), they will lower their utility score from 2 to 1, while if Country B changes from "Action 2" to "Action 1" (Cell D to Cell C), their utility score will fall from 2 to 1. The solution from which neither player will change his/her action selection rationally/logically is called the equilibrium solution (i.e., *Nash Equilibrium*).

However, it is obvious to all players from Table 3.3(b) that the solution combination in Cell A is better than the equilibrium solution in Cell D. The most important issue is how to keep the original Cell A status for all the players without any one player betraying the other player(s). This is the major motivation to set a regime/rule to promote the keeping of the promise.

3.3.3.2 *The Battle of the Sexes (Coordination Game)*

Another regime-type game is *The battle of the Sexes (Coordination Game)*. Suppose that the reader and your partner have a date, and have to choose between a football game (male's preference) or a romantic movie at the theater (female's preference) — which would you choose? If you each of you follow your own preference, the original purpose (dating) will not be realized. This is the original analogical concept behind "the battle of the sexes" game.

Refer to Table 3.3(c). Let us consider an example of the "prisoner's dilemma" game using the case study of the video system standardization battle between Betamax and VHS. The Betamax system was invented by Sony in 1976 and became the first video system standard. However, to reduce the cost and weight of home video systems, JVC developed the VHS system as an alternative — though the Betamax seemed to be of better technological quality. Since the video tapes each system used were also different, the customer had to choose to fully adopt either system exclusively. Video companies hoping to increase their market share without having to produce two versions of their products also needed to choose one unique world standard between Betamax and VHS.

Player A in this 'battle' is Sony, while JVC is Player B; Action 1 is the adoption of VHS as the global standard, and Action 2 establishes Betamax as the world standard. Without reaching an agreement on the global standard, if Sony and JVC produce stronger Betamax and VHS products, respectively (Cell C), they generate a profit score of 2 and 2; while if Sony and JVC produce weaker VHS and Betamax products, respectively (Cell B), their profit scores stand at 1 and 1. In contrast, if either becomes the single global standard, the total profit score increases dramatically to 7. If Betamax is chosen, Sony's profit will be higher (4) than JVC's profit (3) (Cell D), and if the VHS system is chosen, JVC's profit will be 4 and Sony's 3 (Cell A). Note that once the standard is determined, no player has incentive to move to another status. Once the standard was fixed on the VHS system in 1988, Sony could not move back to Betamax (which meant a loss of profit), even though VHS production continued. Sony came in second to JVC.

Let us compare the above two games more closely: generating a regime/rule/ agreement is relatively easy in the "Prisoners' Dilemma," but keeping that status for a long period is difficult. In contrast, generating a regime (setting the global standard) is very tough in "The Battle of the Sexes," but once this is set, no protecting function is necessary. Thus, during the selection process of the standard, even a country's power is occasionally used to provide pressure/threat to opponent countries. Arthur Stein [3] used the terms *Collaboration* and *Coordination* to distinguish between the cooperative actions of the "Prisoners' Dilemma" and the "Battle of the Sexes" [3].

3.3.3.3 *The Stag Hunt Game (Assurance Game)*

The Stag Hunt is another game that may belong to an intermediate game-type that lies between the two games mentioned earlier. See Table 3.3(d). There are two hunters, A and B, and two game, a big stag and a small hare. Stag hunting requires both hunters' collaborative work, while a hare can be caught by just one hunter. If they work collaboratively, they should share a game. Action 1 is "chasing a stag," and Action 2 is "chasing a hare." The payoff will be measured at 4 for half a stag, 3 for one hare, 2 for half a hare, and 1 for nothing. The two players may start from "Action 1"–"Action 1" in Cell A. If these two players continuously chase a stag, they will obtain a stag, leading to the most satisfactory result (4, 4) for both players. If one player changes his mind during the hunt to get a hare instead, he may get one hare, but the other player will get nothing (it is difficult for a single hunter to hunt a stag), leading to the payoff distribution of (3, 1) or (1, 3). If both players change their minds together, they need to share a small hare equally, leading to the payoff (2, 2).

From the payoff table in Table 3.3(d), when Player B chooses "Stag," Player A should choose "Stag" (4), rather than "Hare" (3). While when Player B chooses "Hare," Player A should choose "Hare" (2), rather than "Stag" (1). Thus, Player A does not have the dominant strategy, but there are two Nash equilibrium solutions (Cell A and Cell D), i.e., chasing a stag or chasing a hare collaboratively. As long as both players need to collaborate, it is common sense to choose "Stag" (not a very intriguing game!). However, if there is distrust between the players (i.e., you believe that your partner may betray you) or lack of information (on whether the collaboration will ensure a stag), the collaborative operation on a stag may be terminated. In this sense, the "Stag Hunt" is sometimes called an *Assurance Game*; where the global regime should ferment the *mutual trust* and enhance information transfer to reassure member states of the mutually beneficial status. The regime may not need to force the rules onto the players (there is already a strong incentive to keep the "Action 1"–"Action 1" relation).

To apply this to a real-life situation: the two players are two countries, A and B, who have two action choices, "free trade" and "protective trade." When both countries adopt a mutual free trade policy, both will receive the maximum profit (4, 4). When A keeps free trade and B takes protective trade, A will make a big deficit (1), and B's profit decreases to 3. In this scenario, if A switches to protective trade in reaction, A's deficit will recover to 2 (Cell D). You can imagine that if the "trust" level is mutually low, the two countries' relation will likely move toward the situation in which both exercise protective trade. Once this status is established, because this is a Nash equilibrium status, it is difficult to return to the free trade status without dramatically improving mutual trust.

3.4 Global Regime — Case Studies (Cooperation, Protection, Reduction, Continuation)

Earlier, I proposed the following four-Chinese-character keyword for the era of "politico-engineering," "協, 守, 減, 維 (cooperation, protection, reduction, and continuation)." I also mentioned that International *cooperation* in the standardization of internet systems and computer cables became essential to accelerating mutual communication; that the *protection* of a country's territory and environment from enemies or natural disasters and the spread of infectious diseases is mandatory; that the *reduction in the use* of toxic materials and resources and energy consumption are also key; and that the *continuation* of society, i.e., Sustainable Society, is important to promote. We consider the global regime case studies applicable to these situations, in this section.

3.4.1 *Cooperation*

3.4.1.1 *Ethernet*

Though electrical outlet types differ significantly across countries, computer cables worldwide have already become rather standardized. The Ethernet was developed at Xerox PARC around 1973 [4]. It was inspired by ALOHA net, which *Robert Metcalfe* had studied as part of his PhD dissertation. The idea was first documented in a memo that Metcalfe wrote in 1973, where he named it after the disproven "ether" (traditional hypothetical medium for the propagation of electromagnetic waves). In 1975, Xerox filed a patent application under the name of Metcalfe. Metcalfe left Xerox in June 1979 to form 3Com. He convinced Digital Equipment Corporation (DEC), Intel, and Xerox to work together to promote Ethernet as a standard. The so-called DIX standard, for "Digital/Intel/Xerox," specified 10 Mbit/s Ethernet, with 48-bit destination and source addresses and a global 16-bit Ether-type field, which was published as "The Ethernet, A Local Area Network; Data Link Layer and Physical Layer Specifications."

In 1980, the Institute of Electrical and Electronics Engineers (IEEE) started project 802 to standardize local area networks (LAN). The "DIX-group" submitted the so-called "Blue Book" (CSMA/CD specification) as a candidate for LAN specifications. In addition to CSMA/CD, Token Ring (supported by IBM) and Token Bus (supported by General Motors) were also considered as candidates for the LAN standard. Because the DIX proposal was the most technically complete, and because of the speedy action taken by ECMA, which decisively contributed to the conciliation of opinions within IEEE, the IEEE 802.3 CSMA/CD standard was approved in 1982 [4].

Approval of Ethernet on the international level was achieved by a cross-partisan action working to integrate with the International Electrotechnical Commission Technical Committee (TC83) and the International Organization for Standardization (ISO) Technical Committee 97 Subcommittee 6 (TC97SC6). The ISO/IEEE 802/3 standard was approved in 1984. Figure 3.3 shows an example of an Ethernet connection plug (8P8C modular plug).

3.4.1.2 *Betamax* vs. *VHS*

Let us consider the battle in video system standardization between Betamax and Video Home System (VHS) again from the viewpoint of how VHS won.

Sony had met with Matsushita executives sometime in 1974 to discuss the forthcoming home video market. Both had previously cooperated in the development and marketing of the U-Matic video cassette format. Sony brought along a Betamax prototype for Matsushita's engineers to evaluate, without being

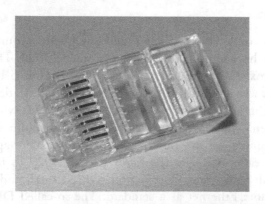

Fig. 3.3.   An 8P8C modular plug for Ethernet systems.

unaware of JVC's work. At a later meeting, Matsushita, with JVC, showed Sony a VHS prototype, and advised them that it was not too late to embrace VHS, but Sony management felt they were too close to production to compromise. Sony had demonstrated a prototype videotape recording system they called "Beta" to the other electronics manufacturers in 1974, and expected that they would back a single format for the good of all. However, JVC had decided to go with its own format, despite Sony's appeal to the Japanese Ministry of Trade and Industry. Thus, the format war began [5].

The main determining factor between Betamax and VHS was the cost of the recorders and their recording time capabilities. Betamax was, in theory, a superior recording format over VHS due to its higher resolution (250 lines vs. 240 lines), slightly superior sound, and more stable images; Betamax recorders were also of higher construction quality. But these differences were negligible to consumers; in their eyes, they did not justify the extra cost of a Betamax VCR (which was often more expensive than a VHS equivalent) or Betamax's shorter recording time.

JVC, which designed the VHS technology, licensed it to any manufacturer who was interested. The manufacturers then competed against each other for sales, resulting in lower prices to the consumer. On the other hand, Sony was the only manufacturer of Betamax for over ten years, and so was not pressured to reduce prices. Sony's decision in 1975 to limit Betamax's maximum recording time to 1 h further handicapped their chances at winning this marketing war. VHS's recording time at first release (1976) was 2 h — meaning that most feature films could be recorded without a tape change.

Although Betamax initially owned 100% of the market in 1975 (as VHS had not launched until the following year), the perceived value of longer recording times eventually tipped the balance in favor of VHS. In Japan, Betamax had more success because of Sony's brand value, but VHS quickly became the market leader after several years. By 1981, sales of Beta machines in the United States had sunk to 25% of the VCR market. As movie studios, video studios, and video rental stores turned away from Betamax, the combination of lower market share and a lack of available titles in Betamax's format further strengthened VHS's position. In the United Kingdom, Beta held a 25% market share, but by 1986 it was down to 7.5%, and continued to decline further.

This VCR format war gave most Japanese engineers (including me) a *psychological shock*. Like Sony's engineers, the engineers believed that "the higher the (quality of) technology, the better the sales," which was similar to Sony's traditional motto: "As long as [Sony] keeps producing top technology, consumers will follow even if the price is higher." Generally, the Japanese (even at present) do not like to purchase cheap items. The collective (psychological) belief is that

products that are "cheaper [are] worse in quality". This basic Japanese corporation strategy (or belief) collapsed as a result of this worldwide marketing war.
International consensus differed significantly from local Japanese consumers'
beliefs. This battle can be considered from the perspective of "The Battle of the
Sexes" game, in which the action choice (Betamax or VHS) is a usual problem,
since both actions are in the Nash equilibrium. Though no official global regime
or standardization effort was made for this VCR format setting, we can also learn
from the JVC/Matsushita team's continuous diplomatic effort on expanding
partnerships with the US RCA and European companies, on top of the VHS's
cheaper production cost.

3.4.2 *Protection*

3.4.2.1 *Treaty on the Non-Proliferation of Nuclear Weapons*

The Treaty on the Non-Proliferation of Nuclear Weapons (NPT) is a multilateral treaty set to prevent the spread of nuclear weapons and weapons technology,
to promote cooperation in the peaceful uses of nuclear energy, and to further
the goal of achieving nuclear disarmament. The Treaty entered into force in
1970 with 190 parties, including the five nuclear-weapon states (US, Russia,
UK, France, and China). *President Barack H. Obama* received the Nobel Peace
Prize in 2009, partially owing to his contribution to this NPT effort. He said in
his Nobel Lecture:

> *"One urgent example is the effort to prevent the spread of nuclear weapons, and to seek
> a world without them. In the middle of the last century, nations agreed to be bound
> by a treaty whose bargain is clear: All will have access to peaceful nuclear power; those
> without nuclear weapons will forsake them; and those with nuclear weapons will
> work towards disarmament … It is a centerpiece of my foreign policy. And I'm work-
> ing with President Medvedev to reduce America and Russia's nuclear stockpiles. But
> it is also incumbent upon all of us to insist that nations like Iran and North Korea do
> not game the system. Those who claim to respect international law cannot avert their
> eyes when those laws are flouted. Those who care for their own security cannot ignore
> the danger of an arms race in the Middle East or East Asia. Those who seek peace
> cannot stand idly by as nations arm themselves for nuclear war …"*

However, critics argue that the NPT cannot stop the proliferation of nuclear
weapons or the motivation to acquire them. They expressed disappointment
with the limited progress on nuclear disarmament; with the five authorized
nuclear weapons states still having 22,000 warhead stockpiles, and having shown
a reluctance to disarm further.

3.4.2.2 *Anti-Personnel Mine Ban Convention 1999*

The *Ottawa Treaty*, an *Anti-Personnel Mine Ban Convention* (a convention on the Prohibition of the Use, Stockpiling, Production and Transfer of Anti-Personnel Mines and on their Destruction) aimed at eliminating anti-personnel landmines (AP-mines) around the world, became binding international law in 1999. To date, there are 162 State Parties to the treaty, while 35 United Nations states, including the United States, Russia, and China are not Party. Engineering developments to find AP-mines will be discussed in Chapter 5 (Crisis Technology).

3.4.3 *Reduction*

3.4.3.1 *Montreal Protocol (on Substances that Deplete the Ozone Layer)*

The Montreal Protocol was ratified in 1989 by 197 parties, which includes 196 states and the European Union. The treaty is structured around several groups of *halogenated hydrocarbons* that play a role in ozone depletion. All these ozone depleting substances contain either chlorine or bromine.

3.4.3.2 *Kyoto Protocol to United Nations Framework Convention on Climate Change*

Negotiations by governments to the United Nations Framework Convention on Climate Change (UNFCCC) on the *Kyoto Protocol* were completed in 1997, committing the industrialized nations to specified, legally-binding reductions in emissions of six *greenhouse gases (GHGs)* — the four GHGs, carbon dioxide (CO_2), methane (CH_4), nitrous oxide (N_2O), sulfur hexafluoride (SF_6); and two groups of gases: hydrofluorocarbons (HFCs) and perfluorocarbons (PFCs) [6]. Thirty-seven industrialized countries and the European Community committed themselves to binding targets for GHG emissions.

It is important to mention the involvement of the United States in this Protocol. *President Bill Clinton* signed the Protocol in 1998. However, the Clinton Administration did not submit the Protocol to the Senate for advice and consent, acknowledging that one condition outlined by S. Res. 98, passed in 1997 — that meaningful participation by developing countries in binding commitments limiting GHGs — had not been met. In 2001, the *George W. Bush* Administration rejected the Kyoto Protocol. The US continued to attend the annual conferences of the parties (COPs) at the UNFCCC, but did not participate in Kyoto Protocol-related negotiations. In 2002, President Bush announced a U.S. policy for climate change that relied on domestic, voluntary actions to reduce the "GHG intensity" (ratio of emissions to economic output) of the U.S. economy by 18% over the next 10 years.

However, as scientific consensus grew that human activities were having a discernible impact on global climate systems (i.e., significantly contributing to a *warming of the Earth* that could result in major impacts such as rising sea levels, changes in weather patterns, and adverse health effects) — and as it became apparent that major nations such as the United States and Japan would not meet their voluntary stabilization target by 2000 — parties to the treaty decided in 1995 that it would be necessary to enter into a legally binding, non-voluntary agreement. Negotiations began on a protocol to establish legally binding limitations or reductions in greenhouse gas emissions. It was decided by the parties that this round of negotiations would establish limitations only for the developed countries (38 nations). This was referred to as the *Berlin Mandate*, which reflected continuation of the principle established in the UNFCCC that parties bore "common but differentiated responsibilities" in dealing with climate change issues, and that the first steps in reducing greenhouse gas emissions should be taken by the 38 developed countries.

3.4.3.3   *Restrictions on the Use of Certain Hazardous Substances*

Restrictions on the use of certain Hazardous Substances (*RoHSs*), short for "Directive on the restriction of the use of certain hazardous substances in electrical and electronic equipment," was adopted in 2003 by the European Union (EU). RoHS applies to all "hazardous" products in the EU, whether they are made within the EU or imported. Though this is a European/local restriction, and does not include the US or Asian countries, most electronic component manufacturers became rather sensitive to the restriction as long as they were exporting products to the EU.

RoHS restricts the use of the following six substances:

- Lead (Pb)
- Mercury (Hg)
- Cadmium (Cd)
- Hexavalent chromium (Cr^{6+})
- Polybrominated biphenyls (PBB)
- Polybrominated diphenyl ether (PBDE)

There are currently over eighty exemptions: (i) copper alloys containing up to 4% lead by weight is permitted; (ii) lead in high melting temperature type solders (i.e., lead-based solder alloys containing 85% by weight or more lead), and (iii) PZT.

Without using the most common PZT, how can the piezoelectric industry survive in the 21st century? Chapter 6 ("Sustainability Technology") will introduce the answer to this question: a recent development in Pb-free piezoelectric ceramics.

3.4.4 *Continuation*

3.4.4.1 *International Whaling Commission*

The international environmental agreement to "provide for the proper conservation of whale stocks and thus make possible the orderly development of the whaling industry," and to implement its economic and environmental goals was signed in 1946. Participation in the *International Whaling Commission* (*IWC*) is not limited to states involved in whaling. As of 2013, there are 88 members. In recent years, the meetings have attracted international media attention due to the growth of the anti-whaling movement.

As the reader should know, according to their respective religious guidelines, traditionally, Hindus do not eat beef, Muslims do not eat pork, and Buddhists do not eat four-legged animals, though they do consume fish — like whales, as they have no-legs — and two-legged fowl like chicken. After World War II, when I was a kid, the only possible inexpensive meats in Japan were whale and chicken. Whales helped keep Japanese citizens fed both during and after World War II. Whale meat made up almost half of all animal protein consumed by the Japanese in the 1950s, and one-quarter in the 1970s.

The Sea Shepherd Conservation Society (SSCS), a marine conservation organization based in Friday Harbor, Washington, USA, was founded in 1977 to protect marine life. In 2008, Animal Planet began filming the weekly series, Whale Wars, based on the group's encounters with the Japanese whaling fleet in the Southern Ocean; a development that brought the group much publicity. The situation appeared to be a matter equivalent to a civil war, with petrol bombs being used against the Japanese research fleets.

In response, Japanese Agriculture Minister Yuji Yamamoto said:

> *"It is whaling research over the last two decades, which paves the way for long-term, sustainable use of this renewable marine food resource. Why not hunt whales if they can be hunted sustainably? And if the principle of sustainable use is compromised on behalf of one animal, what is to stop a domino effect from happening that in time would limit Japan's use of other animal resources that it relies so heavily upon?*

[…] Japanese scientists are also concerned about the possible interplay between the various species of cetaceans in the Antarctic ecosystem arguing that one reason for the slow recovery of the blue whale stocks since the 1960s might be the rapid growth of the Minke whale population."

3.4.4.2 Washington Convention (Convention on Int'l Trade in Endangered Species of Wild Fauna and Flora)

The Convention on International Trade in Endangered Species of Wild Fauna and Flora (CITES) is an international agreement between governments. Its aim is to ensure that international trade in specimens of wild animals and plants do not threaten the survival of the species in the wild, and it accords varying degrees of protection to more than 35,000 species of animals and plants. Japan and Norway proposed to shift some of the whale species from Appendix I (very dangerous) to Appendix II (less dangerous) to relax hunting regulations [7].

Both IWC and CITES regimes were originally set up "for the proper conservation of whales/wild animals." However, the former leans in favor of the anti-whaling movement, and the latter released whale hunting regulations after checking on the animal's conservation status. We can say that Japan and Norway shifted their political "proverbial foot" from IWC to CITES. We call this sort of political action *Forum Shopping*. Forum shopping is the informal name given to the practice adopted by some litigants to have their legal case heard in the court thought most likely to provide a favorable judgment. Some jurisdictions have, for example, become known for being "plaintiff-friendly" and so have attracted litigation even when there is little or no connection between the legal issues and the jurisdiction in which they are to be litigated. It is important to know that the regulating tactic is rather different, depending on each forum or global regime on a similar territory with different member and leadership countries. This is a practical problem with setting up a global regime by collecting various states with different societal, cultural, and psychological milieu.

3.5 Summary

After explaining the levels of societal systematization, we considered global regime setting. We identified three factors to the creation, maintenance, and modification of a regime: *power, profit,* and *norm*. We learned rational decision-making strategies based on *Game Theory*. Six typical 'games' used in Game theory were studied: (1) harmony, (2) independence, (3) zero-sum, (4) *prisoner's dilemma,*

Fig. 3.4. Illustration for six games: (1) harmony, (2) independence, (3) zero-sum, (4) prisoner's dilemma, (5) battle of the sexes, and (6) the stag hunt. Which illustration matches which game?

(5) *battle of the sexes*, and (6) the *stag hunt*. We recognized that the former three types do not create a regime, while the latter three games can. We also established the definition of "politico-engineering" to mean "politically-initiated engineering," i.e., engineering initiated by the necessities of emergencies, by various levels of government — such as municipal, state, country, or union governments — to develop technology to overcome the problem(s). These problems are occasionally related to international relationships — such as wars, global regimes, etc.

A four-keyword slogan for the era of *politico-engineering* — *cooperation, protection, reduction, and continuation* — was proposed. As case studies of global regimes, the following were introduced: The establishment of Ethernet standard as an example of "cooperation"; the Kyoto Protocol and Non-Proliferation Treaty as examples of "protection" and "reduction"; and the concept of "forum shopping" was introduced.

3.6 Problem

Figure 3.4 depicts six typical games: (1) harmony, (2) independence, (3) zero-sum, (4) prisoner's dilemma, (5) battle of the sexes, and (6) the stag hunt. Which pictures (a)–(f) correspond to which games (1)–(6)?

References

[1] Y. Yamamoto, *International Regimes and Global Governance*, Yuhikaku Publ., Tokyo (2008) [in Japanese].

[2] http://www.gr2014parliament.eu/Portals/6/PDFFILES/NA0113090ENC_002.pdf.

[3] A. Stein, Coordination and collaboration: regimes in an anarchic world, *Int. Organ.* **36**(2), 299–324 (1982).

[4] http://en.wikipedia.org/wiki/Ethernet.

[5] https://en.wikipedia.org/wiki/Videotape_format_war.

[6] https://en.wikipedia.org/wiki/Kyoto_Protocol.

[7] https://en.wikipedia.org/wiki/CITES.

Chapter 4

Categorization of Politico-Engineering

I categorize technologies into two types: *crisis technologies* and *normal technologies*. Crisis technologies include technologies meant to prevent, resolve, mitigate, or recover from (1) natural disasters, (2) infectious diseases, (3) enormous accidents, (4) terrorist/criminal incidents, and (5) wars/territorial invasions. Normal technologies are further categorized into three sub-categories: (1) new discoveries/inventions, (2) technologies for designing/manufacturing/ marketing, and (3) sustainability technologies. In my proposition, "Politico-Engineering" covers (1) crisis technologies, and (2) legally regulated normal technologies such as sustainability technologies.

Crises are generated by external actors suddenly/unexpectedly, and as such, crisis technologies are usually urgently required to solve the problem particular to that moment, as quickly as possible. In contrast, sustainability technologies are required to solve long-term problems that may threaten societal continuation, and can be planned and adopted over a relatively longer period. This chapter provides an overview of the current "politico-engineering" (politically-initiated engineering) technologies for crises and societal sustainability. The detailed technological and engineering developments are described in the successive chapters, Chapters 5 and 6.

4.1 Crisis Technologies

Crisis technologies can be further classified into (technologies for):

- Natural disasters (earthquakes, volcanic eruptions, tsunamis, tornadoes, hurricanes, floods, lightning, etc.);
- Epidemic/infectious diseases (e.g., smallpox, polio, measles, HIV, EBOLA, etc.);
- Enormous accidents (e.g., the Three Mile Island core meltdown, BP oil spill, etc.);
- Intentional accidents (e.g., acts of terrorism, criminal activity, cyber threats, etc.);
- Civil wars, international wars, territorial aggression.

4.1.1 *Natural Disasters*

A natural disaster is a major adverse event resulting from natural processes of the earth. Examples include earthquakes, volcanic eruptions, tsunamis, tornadoes, hurricanes, floods, and lightning. A natural disaster can cause the loss of life or property damage, and typically leaves some economic damage in its wake — the severity of which depends on the affected population's resilience, or ability to recover.

In 2012 (a moderate year), there were 905 natural disasters worldwide, 93% of which were weather-related disasters. Overall costs were US$170 billion and insured losses of $70 billion. Of these, 45% were *meteorological* disasters (storms), 36% were *hydrological* (floods), 12% were *climatological* (heat waves, cold waves, droughts, wildfires) and 7% were *geophysical* events (earthquakes and volcanic eruptions), according to the report by Worldwatch Institute [1].

Between 1980 and 2011, geophysical events accounted for 14% of all natural catastrophes, including the Kanto–Tohoku Big Earthquake and the resulting Tsunami on March 11, 2011 (Fig. 4.1). This incident disclosed that Japan does not have sufficient earthquake prediction technologies, tsunami warning systems, as well as unmanned aerial and underwater surveillance vehicles for finding bodies and debris, which significantly increased the death toll.

4.1.2 *Epidemics/Infectious Diseases*

An epidemic is the rapid spread of an *infectious disease* to a large number of people in a given population within a short period of time — usually two weeks or less.

Fig. 4.1.   Tsunami attack on Sendai area after the Kanto–Tohoku Big Earthquake on March 11, 2011 (picture is a screenshot of a news broadcast).

For example, within a year from the first recorded case of Ebola on March 23, 2014, more than 10,460 people had been reported to have died from Ebola in the following six countries: Liberia, Guinea, Sierra Leone, Nigeria, the US, and Mali, which really makes it a *pandemic* [2].

Epidemics of infectious disease are generally caused by several factors including a change in the ecology of the host population (e.g., increased stress or an increase in the population density of a vector species) and a genetic change in the pathogen reservoir, or the introduction of an emerging pathogen to a host population (by movement of the pathogen or host). Generally, an epidemic occurs when the host's immunity to either an established pathogen or a newly emerging/ novel pathogen is suddenly reduced below that found in the endemic equilibrium and the transmission threshold is exceeded.

An epidemic may be restricted to one location; however, if it spreads to other countries or continents and affects a substantial number of people, it may be termed a *pandemic*. The declaration of an epidemic usually requires a good understanding of a baseline rate of incidence. For example, a few cases of a very rare disease may be classified as an epidemic, while many cases of a common disease (such as the common cold) would not. Certain diseases, such as influenza, are defined as epidemics only when they reach some defined increase in incidence above their baseline.

4.1.3 *Enormous Accidents*

An 'enormous accident' is defined in this book as a significant accident caused *unintentionally* by a human or human activities. This is exemplified by the Three Mile Island nuclear power plant meltdown in 1979 and the BP Deepwater Horizon oil spill in 2010.

4.1.3.1 *Three Mile Island Nuclear Power Plant Meltdown*

The *Three Mile Island accident* was the first nuclear meltdown in 1979, in one of the two nuclear reactors in Pennsylvania, USA, one year after I first joined Penn State University as a research associate. The incident was rated "5" on the seven-point International Nuclear Event Scale. The accident began with failures in the non-nuclear secondary system, followed by a stuck-open pilot-operated relief valve for nuclear reactor coolant in the primary system. The mechanical failures were compounded by inadequate training and human factors. The partial meltdown resulted in the release of unknown amounts of radioactive gases and radioactive iodine into the environment [3]. Cleanup started in August 1979, and officially ended in December 1993, with a total cleanup cost of about US$1 billion [4].

4.1.3.2 *Deepwater Horizon Oil Spill*

Though the cause of the explosion on the rig is not clear, the *Deepwater Horizon oil spill* in the Gulf of Mexico began on April 20, 2010, on the BP-owned Transocean-operated Macondo Prospect [5].

"BP is responsible for this leak — BP will be paying the bill," *President Barack Obama* told reporters after the incident. BP said that it was ready to pay all legitimate claims tied to the oil spill caused by the accident. But the head of BP Group *Tony Hayward* said it could have been prevented if Transocean's blow-out preventer (BOP) had done its job and prevented the explosion. A BOP is a large valve at the top of a well, and activating it will stop the flow of oil. Hayward told CNN, "[The BOP] is the ultimate fail-safe mechanism … And for whatever reason — and we don't understand that yet, but we clearly will as a consequence of both our investigation and federal investigations — the BOP by Transocean failed to operate."

The US Government estimated the total discharge at 4.9 million barrels (780,000 m^3). *Containment booms* stretching over 1,300 km were deployed, either to corral the oil or as barriers to protect marshes, mangroves, shrimp/crab/oyster ranches or other ecologically sensitive areas. Booms extend 0.46–1.22 m above and below the water surface and were effective only in relatively calm and slow-moving waters. Including one-time use sorbent booms, a total of 4,100 km of booms were deployed. The Louisiana barrier island plan was developed to construct barrier islands to protect the coast of Louisiana. Critics allege that the decision to pursue the project was political with little scientific input, and the Environmental Protection Agency (EPA) also expressed concern that the berms would threaten wildlife.

4.1.4 *Intentional Accidents*

Intentional accidents cover incidents of intentional wrongdoing such as *acts of terrorism*, criminal activities, and intentional aircraft crashes/shipwrecks, as exemplified by the World Trade Center attack in 2001 and the Germanwings crash in March 2015.

4.1.4.1 *World Trade Center Attack*

The September 11 attacks were a series of four coordinated terrorist attacks by terrorist group *al-Qaeda* on the US in New York City and the Washington, D.C. metropolitan area, on Tuesday, September 11, 2001. The attacks killed 2,996 people and caused at least $10 billion in property and infrastructure damage [6].

Fig. 4.2.   Terrorist-hijacked planes hitting the World Trade Center towers on September 11, 2001.

Four passenger airliners were hijacked by 19 al-Qaeda terrorists to be flown into buildings in suicide attacks. Two of the planes, American Airlines Flight 11 and United Airlines Flight 175, were crashed into the North and South towers of the *World Trade Center complex* in New York City. Within 2 h, both 110-story towers collapsed and the resulting fires caused the partial or complete collapse of all the other buildings in the WTC complex, including the 47-story World Trade Center tower, as well as significant damage to 10 other large surrounding structures (Fig. 4.2). A third plane, American Airlines Flight 77, was crashed into the *Pentagon* (HQ of the United States Department of Defense), leading to a partial collapse on its western side. The fourth plane, United Airlines Flight 93, was targeted at Washington, D.C., but crashed into a field near Shanksville, Pennsylvania, after its passengers tried to overcome the hijackers. In total, 2,996 people died in the attacks, including the 227 civilians and 19 hijackers aboard the four planes. It was the deadliest incident for firefighters and law enforcement officers in the history of the United States, with 343 and 72 killed respectively.

4.1.4.2   *Germanwings Crash*

A Germanwings Airbus A320 crashed on March 24, 2015 near Seyne-les-Alpes, in the French Alps [7]. Authorities said that 27-year-old co-pilot *Andreas Lubitz*, who in the past had been treated for *suicidal tendencies*, locked his captain out of the cockpit before deliberately crashing the Airbus 320 into a mountain in the French Alps. All 150 people aboard Flight 9525 from Barcelona to Duesseldorf were killed.

The airline said that as part of its internal research they found e-mails that Lubitz sent to the Lufthansa flight school in Bremen when he resumed his training there after an interruption of several months. In them, he informed the school that he had suffered a "serious depressive episode," which had since subsided. Lubitz subsequently passed all medical checks. That airline claimed to have provided the documents to prosecutors, and declined to make any further comment. The revelation that Lufthansa officials had been informed of Lubitz's psychological problems raises further questions about why he was allowed to become a pilot for its subsidiary, Germanwings, in September 2013. At the time of writing, the investigation was still ongoing.

4.1.4.3 *Cyber Threats*

While the reader has probably read about the security breaches at Sony Pictures Entertainment by North Korea in 2014, you might not know about the hacking that caused massive physical damage to a steel mill in Germany [8]. The hackers had manipulated and disrupted the systems that control the mill to such a degree that a blast furnace could not be properly shut down. Someone gained unauthorized access via the Internet to Russia's Sochi arena's heating and cooling system, as well as its emergency response system. The system was fortunately reconfigured in time for the start of the 2014 Winter Olympics and its opening ceremonies. Industrial control systems monitor those critical, yet ordinary processes that run nearly every large operation. Manipulating them means that hackers could turn off electricity, stop the flow of clean water, shut down transit systems, cause equipment damage in the millions of dollars, and even kill people.

A report issued in 2015 by the US Department for Homeland Security says that in 2014, the Industrial Control Systems Cyber Emergency Response Team reacted to 245 such incidents. The energy sector reported the highest number of incidents that year with 79, followed by manufacturing at 65, and health care at 15.

IEEE Senior Member *Joseph Weiss*, a cyber security expert, wrote:

"The experts you have brought together are focused on the security and privacy of IT. Unfortunately, that is only part of the cybersecurity landscape. There is still a significant gap in understanding about control system cyber-security by the traditional IT security community. There have already been almost 400 control system

cyber incidents but very few were identified as such. Once control systems go down, protecting credit card information will be the least of our worries."

In Weiss' lectures on control system risk and control system cyber forensics at the International Atomic Energy Agency, the US Air Force Institute of Technology, and Stanford University [8], he said:

"The difference between the IT world and the control systems world is that IT is only worried about a malicious attack. In our world, a nonintentional attack kills people. If you can do something unintentionally, you can do it worse intentionally. These systems were not meant to be used in the wrong way. It is really difficult for the defensive side to anticipate what the offensive world threats will be when we don't think that way."

4.1.5 *War and Territorial Aggression*

War, civil-war and territorial aggression are considered in this subsection.

4.1.5.1 *Vietnam War*

The *Vietnam War* was a Cold War-era proxy war that occurred in Vietnam, Laos, and Cambodia from November 1, 1955 to the fall of Saigon on April 30, 1975 [9]. This war followed the First Indochina War (1946–1954) and was fought between North Vietnam (supported by the former Soviet Union, China, and other communist allies) and the government of South Vietnam (supported by the US and other anti-communist allies). The Viet Cong (also known as the National Liberation Front, or NLF), a South Vietnamese communist common front aided by the North, fought a guerrilla war against anti-communist forces in the region. The People's Army of Vietnam (also known as the Vietnamese People's Army — the North Vietnamese Army, or NVA) engaged in a more conventional war, at times committing large units to the battle.

As the war continued, the part of the Viet Cong in the fighting decreased as the role of the NVA grew. The US and South Vietnamese forces relied on air superiority and overwhelming firepower to conduct search and destroy operations, involving ground forces, artillery, and airstrikes. In the course of the war, the US conducted a large-scale strategic bombing campaign against North Vietnam, and over time, the North Vietnamese airspace became the most heavily defended in the world. The US government viewed American involvement in the war as a way to prevent a Communist takeover of South Vietnam.

According to the *American Domino Theory*, if one state went Communist, other states in the region would follow, and US policy thus held that Communist rule over all of Vietnam was unacceptable. The domino theory suggested by former *President Dwight D. Eisenhower* was used by successive US administrations during the Cold War to justify the need for American intervention around the world.

Beginning in 1950, American military advisors arrived in what was then French Indochina. The US involvement escalated in the early 1960s, and regular US combat units were deployed beginning in 1965. Operations crossed international borders — bordering areas of Laos and Cambodia were heavily bombed by US forces as American involvement in the war peaked in 1968. The US gradually withdrew their ground forces as part of a policy by *Richard Nixon* known as "*Vietnamization*," which aimed to end American involvement in the war while transferring the task of fighting the Communists to the South Vietnamese themselves. Despite the *Paris Peace Accords*, which was signed by all parties in 1973, the fighting continued. In the US and the Western world, a large anti-Vietnam War movement developed. This movement was part of a larger counterculture of the 1960s. The capture of Saigon by the North Vietnamese Army in April 1975 marked the end of the war, and North and South Vietnam were reunified in the following year. The war exacted a huge human cost in terms of fatalities: Estimates of the number of Vietnamese service members and civilians killed vary from 800,000 to 3.1 million. Some 200,000–300,000 Cambodians, 20,000–200,000 Laotians, and 58,220 US service members also died in the conflict.

4.1.5.2 *Senkaku Islands Conflict*

An incident involving a Chinese fishing trawler and two Japanese Coast Guard patrol boats in Japan's exclusive economic zone near the Senkaku Islands occurred on September 9, 2010. The area has been a part of a territorial dispute over a group of uninhabited islands known as the Senkaku Islands in Japan, Diaoyu in the People's Republic of China (PRC), and the Tiaoyutai Islands in the Republic of China (ROC or Taiwan).

Prior to the 1970s, neither the PRC nor ROC governments made any official statements disputing or claiming sovereignty over the Senkaku Islands. Several maps, newspaper articles, and government documents from both countries after 1945 refer to the islands by their Japanese name, while some even explicitly recognize their status as Japanese territory. *The People's Daily*, a daily newspaper of the Central Committee of the Communist Party of China, in an article published on January 8, 1953, referred to the Senkaku Islands by the Japanese name "Senkaku Shotō" and described the islands as a part of then-US-occupied Ryukyu Islands, which consist of seven groups of islands: Senkaku Islands, Sakishima Islands, Daito Islands, Okinawa Islands, Oshima Islands, Tokara Islands, and Osumi Islands. There have been many official maps published by both governments after 1945 that recognized the islands as Japanese territory [10].

Although Chinese authorities did not assert claims over the islands while they were under US administration, formal claims were announced in 1971 when the US was preparing to end its administration. It might be good to point out that a 1968 academic survey undertaken by the United Nations Economic Council for Asia and the Far East found possible *oil reserves* in the Senkaku area; which may explain the emergence of the recent Chinese claim. The Chinese Ministry of Defense responded after the September 2010 conflict that:

"If Chinese drones entered what Japan considered its territory, Japan might shoot them down, by declaring that China would consider such an action an 'act of war'."

State-controlled media in China warned that:

"China's comprehensive military power ... is stronger than Japan's."

Figure 4.3 shows the number of Chinese vessels entering the territorial waters near the Senkaku Islands after 2010 till 2015. In parallel, the number of times the Japan Air Self-Defense Force was scrambled against Chinese aircraft also drastically increased within the same timeframe (e.g., the Japan Air Self-Defense Force had been scrambled ~500 times per year in 2014).

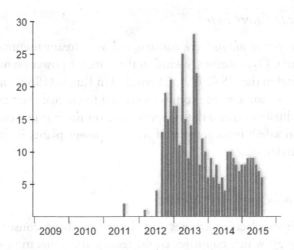

Fig. 4.3.   Number of Chinese vessels entering the territorial waters near the Senkaku Islands.

4.2 Sustainability Technologies

Sustainability technologies are a type of 'normal' technology (i.e., technology which can either be developed by politically-initiated motives or legally regulated). They include:

- Power and energy (e.g., to counter the lack of oil, improve nuclear power plants, or discover new forms of energy/energy harvesting)
- Rare material (e.g., rare-earth metals, Lithium, etc.)
- Food and Biofuel (e.g., rice, corn)
- Toxic materials

  o Restriction (of heavy metals, e.g., Pb, Dioxin)
  o Elimination/neutralization (e.g., mercury, asbestos)
  o Replacement material

- Environmental pollution
- Energy efficiency
- Bio/medical engineering

4.2.1 *Power and Energy*

According to the general reduction or lack of fossil energy resources such as oil, coal, and natural gas, and also owing to CO_2 generation restrictions introduced to counter/slow down "Global Warming" or the "Greenhouse Effect", an energy resource shift is aggressively promoted.

4.2.1.1 *Nuclear Power Plant*

Though *nuclear power plants* were constructed as a dream resource after World War II to reduce CO_2 emissions, owing to three major power plant accidents — Three Mile Island in the US (1979), Chernobyl in Russia (1986) and Fukushima in Japan (2011) — some of the countries started to change their energy policies in order to gradually decrease their dependence on nuclear power. A significant improvement in safety management of nuclear power plants is an urgent issue for all the countries.

4.2.1.2 *Hydroelectric Power Station*

Conventional hydroelectric turbines and dams have been considered as possible means through which countries could reduce the generation of CO_2. This

Fig. 4.4.   The Three Gorges Dam in China, the largest hydroelectric power station in the world.

opinion led to the building of the 22,500 MW *Three Gorges Dam* in the People's Republic of China, the largest hydroelectric power station in the world, in 2012 (Fig. 4.4). However, the dam flooded archaeological and cultural sites and displaced some 1.3 million people, and is causing significant ecological changes, including an increased risk of landslides. The dam has been a controversial topic both domestically and abroad [11].

4.2.1.3 *Renewable Energy*

In parallel, *renewable energy* — such as solar cells, wind turbines, geothermal and tidal wave energy harvesting systems — is a hot research topic. It will be further discussed in Chapter 6.

4.2.2 *Rare Materials*

During my stay in Japan as a "Navy S&T Ambassador to Japan," I experienced weak Japanese government policies on "rare materials", in particular, in the superior technology areas of electronic components: super magnets with rare earth materials and lithium-ion batteries (LIBs).

4.2.2.1 *Rare Earth Metals*

Though most of the production of bonded magnets has been moved to China, Japan is still the leader in innovation for magnets and holds most of the patents for new high-energy materials — particularly in *rare earth magnets* such as *neodymium*. In contrast, China is by far the leader for all magnet materials. This occurred after the middle of 1990s. China mines over 55% of the world's rare earth ores needed for most magnets (there is no rare earth mine in Japan). This, coupled with low labor costs and few health and environment regulations, has enabled them to grow their magnet industry at 70–130% per year.

As introduced in the previous section, on September 9, 2010, a Chinese fishing trawler attacked Japanese Coast Guard patrol boats in the *Japanese exclusive economic zone,* near the *Senkaku* Islands — a territory close to key shipping lanes and rich fishing grounds, and in which oil and "methane hydrate" reserves were recently discovered. Aside from a 1945 to 1972 period of administration by the United States as part of the Okinawa Islands, this archipelago was officially returned to Japan in 1971 under the Okinawa Reversion Agreement between the US and Japan. However, since around the time energy reserves were found in the area, acts seen by Japan as aggressive territorial invasion began being committed by China (and continue to be committed today). After the collisions, the Japanese Coast Guard arrested the Chinese trawler captain, who was held by the Japanese Coast Guard until September 24, 2010; after which, the Japanese government released this captain without imposing any penalty [12]. Though the reason for this was not clearly mentioned by the then-Japanese cabinet, major news and media channels reported that "under the pressure by the rare-earth metal export regulation against Japan, the Chinese government pushed Japan to release the captain." For the reader's information, China's Ministry of Commerce eliminated the above-mentioned "quota system" for rare-earths exports starting in January 2015. China had used the rare-earth metal exportation as a trade-off negotiation tool for releasing the trawler captain without any penalty.

4.2.2.2 *Lithium*

A lithium-ion battery (LIB) is a member of a family of rechargeable battery types in which lithium ions move from the negative electrode to the positive electrode during discharge and back when charging. Japan is a leading country in the production of high energy density, safe and reliable LIBs. The Nissan Leaf is an example of a compact five-door hatchback electric car manufactured by Nissan and introduced in Japan and the United States in December 2010 that makes use of lithium-ion battery technology. However, half the world's known lithium reserves are located in Bolivia, a nation sitting along the central eastern slope of the Andes. In 2009, Bolivia was negotiating with Japanese, French, and Korean firms to begin extraction. The former president of South Korea, *Lee Myung-bak*, successfully inked the "Lithium Deal" with Bolivia to mine the world's largest salt flats in 2011. *The Korea Herald* reported that South Korea's bid would tap on the largest lithium deposits in the world. Korea, the chaser of Japanese lithium battery technology, may obtain advantages over Japan in rechargeable batteries for mobile phones, laptops, and electric cars, through this deal because Japan will lose their lithium import.

4.2.2.3 *Oil and Civil War*

The reduction in or a new discovery of natural resources such as oil, are occasionally related to incidents of civil warfare and/or a *coup d'état* in various resource countries, such as Angola, Sudan, Nigeria, etc.

One such event is the *In Amenas hostage crisis*, which began on January 16, 2013. Al-Qaeda-linked terrorists affiliated with a brigade led by *Mokhtar Belmokhtar* took expatriate hostages at the Tigantourine gas facility near In Amenas, Algeria. One of Belmokhtar's senior lieutenants, *Abdul al Nigeri*, led the attack and was killed. After four days, the Algerian Special Forces raided the site, in an effort to free the hostages. At least 39 foreign hostages from nine different countries (of which 10 were Japanese) were killed [13]. BP (former British Petroleum) is being sued in the USA and in the UK for failing to properly protect their staff. This incident will be reexamined in Chapter 7 (Risk Management).

4.2.3 *Food*

4.2.3.1 *Lack of Food*

A *famine* is defined as a widespread scarcity of food caused by several factors, including crop failure, population imbalance, or government policies. This phenomenon is usually accompanied or followed by regional malnutrition, starvation, an epidemic, and increased mortality. Nearly every continent in the world has experienced a period of famine throughout history. Some countries, particularly in Sub-Saharan Africa, continue to have extreme cases of famine. To solve famine problems, genetic bio-engineering studies aimed at creating durable crops for severe weather conditions have been employed.

4.2.3.2 *Biofuel*

Rice and corn are, of course, foods, but they are also bio-materials for producing *biofuel*. At the end of 1990s, the rice export price in Thailand drastically increased by more than 60% because of imports of rice by the US and Japan as biofuel materials for alcohol production. Due to this increase in exports and the significant increase in the price of rice, it became difficult for Thai people in the lower income-tiers to purchase rice, i.e., a temporary famine.

4.2.4 *Toxic Materials*

Toxic waste is a waste material that can cause death, injury, diseases, or birth defects in living creatures. Discarded materials can pose a long-term risk to human and animal health or the environment. The term is often used interchangeably with "hazardous waste". Hazardous wastes are poisonous by-products of manufacturing, farming, city septic systems, construction, automotive garages, laboratories, hospitals, and other industries. The waste may be liquid, solid, or sludge, and contain chemicals, heavy metals, radiation, dangerous pathogens, or other toxins. Even households generate hazardous waste from items such as batteries, used computer equipment, and leftover paints or pesticides.

4.2.4.1 *Minamata Disease*

Minamata Bay, in the southern part of Japan, was heavily polluted in the 1950s and 1960s by wastewater mixed with *mercury* (methylmercury), dumped by the Nippon Chisso Corporation's factory (Fig. 4.5). The highly toxic compound bio-accumulated in fish and shellfish in the bay — which, when eaten by the people living around the bay, gave rise to *Minamata Disease*, a neurological syndrome whose symptoms included insanity, paralysis, coma, and death. More than 10,000 people were affected [14].

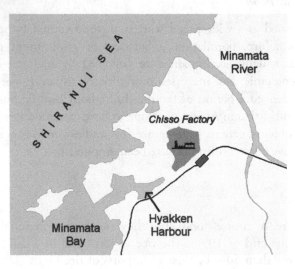

Fig. 4.5.   The Chisso factory and its wastewater routes.

4.2.4.2 *Environmental Protection Agency*

In the United States, the Environmental Protection Agency (EPA) and the state departments oversee the rules that regulate hazardous waste. The EPA requires toxic waste to be handled with special precautions and be disposed of in designated facilities around the country. Also, many cities in the United States have collection days when household toxic wastes — including some materials that may not be accepted at regular landfills, ammunition, commercially-generated waste, explosives/shock-sensitive items, hypodermic needles/syringes, medical waste, radioactive materials, and smoke detectors — are gathered.

4.2.4.3 *Restrictions of Hazardous Substance*

As already introduced in Section 3.4.3.3, the RoHS (Directive on the restriction of the use of certain hazardous substances in electrical and electronic equipment) came into effect in the EU, in 2003 [15]. RoHS applies to all products in the EU, which contain any amount of combination of six hazardous materials, regardless of whether they are made within the EU or imported.

4.2.5 *Environmental Pollution*

Environmental pollution is the introduction of contaminants into the natural environment that cause adverse change. *Pollution* can take the form of chemical substances, noise, heat, or light. Pollutants, the components of pollution, can be either foreign substances/energies or naturally occurring contaminants.

Particulate matter (*PM*) are microscopic solid or liquid matter suspended in the Earth's atmosphere. The term *aerosol* commonly refers to the particulate/air mixture, as opposed to the particulate matter alone. Sources of particulate matter can be man-made or natural. They have impacts on climate and precipitation that adversely affect human health. Subtypes of PM include (1) suspended particulate matter (SPM), (2) respirable suspended particles (RSP; particles with diameters of 10 μm or less), and (3) fine particles (diameter of 2.5 μm or less) [16].

Chinese cities have an air pollution problem, which has become an increasingly hot topic both inside and outside China's borders (see Fig. 4.6). Most of China's polluted cities have and continue to reach *Air Quality Index* (*AQI*) levels higher than 500, and have even crossed 1,000 (for comparison, the United States considers any AQI higher than 150 to be unhealthy, and any exceeding 300 to be hazardous). China's air pollution is caused in part by its traffic and heavy industries, but it is worst in winter, when cities are heated, in a large part, by power generated by the burning of coal. This generates large quantities of airborne particles (*PM 2.5*), and when winds and precipitation are low, these

Fig. 4.6. PM 2.5 suspended in the atmosphere.

particles can remain in the air over cities for extended periods of time. This situation can be exacerbated if, like Beijing, a city is surrounded on several sides by mountains that serve to "trap" the air pollution [16].

Another sort of environmental pollution is global warming. Global warming originated from the increase of *greenhouse gases* (*GHGs*), such as carbon dioxide (CO_2), methane (CH_4), nitrous oxide (N_2O), and sulfur hexafluoride (SF_6), and is forcing ice shelves to calve, producing monolith icebergs jutting into the Antarctic Ocean.

4.2.6 *Energy Efficiency*

Achieving energy efficiency means reducing the amount of energy required to provide products and services. For example, insulating a home allows a building to use less energy for heating and cooling to achieve and maintain a comfortable temperature. Installing fluorescent lights or natural skylights reduces the amount of energy required to attain the same level of illumination compared to using traditional incandescent light bulbs. Compact fluorescent lights use one-third the energy of incandescent lights and may last from 6 to 10 times longer. Improvements in energy efficiency are generally achieved by adopting more efficient technology or production processes, or by applying commonly accepted methods to reduce energy loss. Reducing energy use is also seen as a solution to the problem of reducing carbon dioxide emissions. According to the International Energy Agency, improved energy efficiency in buildings, industrial processes, and transportation could reduce worldwide energy needs in 2050 by 1/3, and help control global emissions of GHGs.

Energy efficiency and *renewable energy* are said to be the twin pillars of sustainable energy policies and are high priorities in the sustainable energy hierarchy. In many countries, energy efficiency is also perceived to have a national security benefit because it can be used to reduce the level of energy imports from foreign countries and may slow down the rate at which domestic energy resources are depleted.

4.2.7 *Bio/Medical Engineering*

The fourth keyword, "continuation," in the slogan for the politico-engineering period, has to do with societal continuation or the maintenance of the *status quo*; i.e., "sustainability" in terms of humanity, nature, and resources. Resource consumption is balanced by resources assimilated by the ecosystem, including food, fiber, water, and energy. Genetic engineering is highly focused on increasing food production. In parallel, medical research elongates the human lifetime.

Japan has retained the top spot in terms of its citizens' *average life expectancy* for both sexes: men and women live to an average age of 81 and 87, respectively (the combined average is 84 years), according to statistics by the *World Health Organization* (WHO). Japan has placed first in the annually compiled list for more than 20 years, but more countries are likely to catch up with Japan, where the smoking rate is high, a WHO official said.

An increasing lifespan (mainly due to the improvements in medical and pharmaceutical technology) seems to be welcomed on the level of the individual. However, it is notable that the Japanese pension system is gradually approaching bankruptcy because of the combined effect of longer lifetimes and low birth rates. I know this as I am a recipient of the *Japanese pension* because I contributed to the pension system for 25 years in Japan, until I changed my citizenship many years ago (I am American now). In my (and my wife's) case, the amount received is gradually decreasing year by year. In addition, the new mandatory Medicare insurance premium started to be deducted from this amount since 2013, even if the recipient does not reside in Japan (it appears to be a government strategy to reduce the pension paid). According to my own simple calculations (from the time I received my first pension payment, to the day I die), only less than half of what I had originally contributed to the pension fund during my 25-year working period in Japan will ever be paid to me. In other words, my previous pension deposit has already been spent on covering my parents' generation. The present status is further creating a more serious negative-spiral (in fact, all my children have decided to opt out of the national pension, since they know that their contributions would pay for my generation's current pension rather than their own in the future). Without changing the current pension system (i.e., where the present pension reduction from a worker's salary funds the coverage for the previous generation's, and not the worker's own future payment), the Japanese pension system is likely to reach bankruptcy in 15–30 years if the senior population continues to expand at the same rate as at present. Of course, I am not suggesting that we decelerate the development of medical technology in order to stop further extension of the average human life span. The opinion voiced only describes the serious governmental/societal problem in Japan that has arisen because of the longer average expected lifespan.

4.3 Paradigm Shift of Crisis Technology Development

Finally, we compare the difference between the *product planning strategies* for techno-, econo-, and politico-engineering. As illustrated in Fig. 4.7, in *techno-engineering*, as seen in the example of single crystal relaxors or piezoelectrics, the fundamental Science and Technology (S&T) or the *discovery* (i.e., the key technology) comes first. Because I am the discoverer of this gigantic electro-mechanical coupling factor (that more than 90% of the input electric energy can be converted into output mechanical energy), the US Navy recruited me from a Japanese University to Penn State University. Then, significant investment was put into the rapid development of underwater sonars, hydrophones,

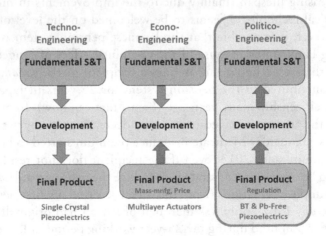

Fig. 4.7. Difference among the product planning strategies for techno-, econo-, and politico-engineering.

sound projectors, etc. Finally, application products were created with searching off-shore materials (i.e., single crystal) suppliers in parallel.

In the case of econo-engineering, the piezoelectric multilayer component is a good example of the trend's tendency to have the final specs of the technologies' mass-production capability and manufacturing cost, coming first.

Based on these desired specs, the manufacturing processes are developed as the key technology. My invention of the fundamental design of multilayer actuators was one such starting point in the late 1970s.

In contrast, in politico-engineering, legal regulations for the final products come first, with strict constraints in terms of law-issuing dates, and performance regulations: For example, for regulations determmining the permitted amount of Pb in the final product, the CO_2, NO_x, or SO_x emission amount, specifications for rescue robots, or new weapons. For case studies of politico-engineering driven developments, we will consider the automobile diesel injection valve and deformable mirror.

4.3.1 *Automobile Diesel Injection Valve*

In the 1980s, most of Japanese automobile customers sought vehicles with the highest performance, i.e., high acceleration and speed, rather than high efficiency or mileage, as well as gasoline of the highest octane (often higher than 100, or sometimes, 110). However, in the 21st century, partly because of the worldwide economic recession and increase in oil prices, and partly because of the *exhaust gas regulation*, the automobile engineer's major task has shifted toward developing highly efficient vehicles, with high gas mileage and low CO_2 emissions.

But is a vehicle with a low CO_2 generation rate really a better choice? Diesel engines rather than regular gasoline cars are actually the recommended option from the energy conservation and global warming viewpoint. Why? To understand this, we need to consider the total energy of gasoline production — from *well-to-tank and tank-to-wheel*. Energy efficiency, measured by the total energy required to realize one unit of driving distance for a vehicle (MJ/km), is of course better for high octane gasoline than diesel oil. However, since the electric energy required for the purification process is significant, gasoline's efficiency becomes inferior to diesel's. However, as it is well known, the conventional diesel engine generates toxic exhaust gases such as SO_x and NO_x. In order to solve this problem, new diesel injection valves were developed by Siemens, Bosch, and Toyota with piezoelectric multilayer actuators as the key component. The valve's technical details are discussed in Section 6.5, "Environmental Pollution." This is one of the best examples of how the engineers need to develop devices/systems according to global regimes such as the Kyoto Protocol.

4.3.2 *Deformable Mirror*

Precise wave front control with as small a number of parameters as possible and of compact construction is a common and basic requirement for adaptive optical systems. Continuous surface *deformable mirrors*, for example, tend to be more desirable than segmented mirrors in terms of controllability. In the 1970s during the Cold War, in order to monitor the former Soviet Union's military bases, the use of a deformable mirror was proposed to correct the distortion that occurs in a satellite telescope's image due to atmospheric conditions — as schematically illustrated in Fig. 4.8. However, the original monolithic piezoelectric deformable mirror required many electrodes with many electrical leads to the individual electrode elements, making the total weight and volume a serious problem for installation on satellites. Thus, our Penn State group proposed a much simpler multimorph deformable mirror, which can be more simply controlled by a micro-computer. The technical details of this mirror are discussed in Section 5.5, "Civil War, War, Territorial Aggression." I will introduce, here, the development methodology of this piece of crisis technology: (1) final product specs constraints, (2) development acceleration methodology, and (3) fundamental materials search.

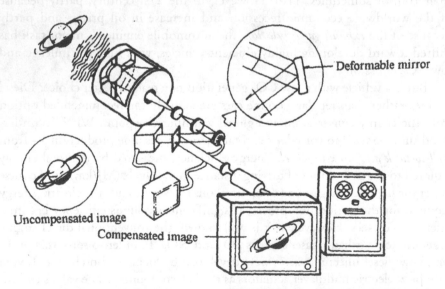

Fig. 4.8.   A telescope image correction system using a monolithic piezoelectric deformable mirror.

Our three distinguished innovations include:

(1) A simple, thin, and light-weight mirror component design for satellite installation;
(2) The Application of computer Finite Element Method simulation on the piezoelectric devices to accelerate design optimization for the first time in the world; and
(3) Invention of new lead-magnesium-niobate (PMN) electrostrostrictive materials, which could eliminate hysteresis significantly in the deformation contour of the mirror, in comparison with conventional piezoelectric PZT materials, and operation capability at harsh/low temperatures in space.

Now regarding the development acceleration methodology — we consider here why so-called "big science" is so advanced in the USA as compared to Japan. I discuss this issue from the viewpoint of a management structure.

In 1978, while at Penn State University, I worked on a NASA contract researching "electrostrictive actuators for precise positioning of optical lenses and mirrors." After spending a year on the project, I was invited to the NASA Workshop on this "Adaptive Mirror Project" (it was named the *Hubble Telescope* when it was launched on the Space Shuttle *Discovery* in 1990). I was surprised to learn that the same research topic I had been working on was also being done independently in an American company (then Honeywell Laboratory). Through the annual workshop, the better results were obtained, and that set a new starting point for next year's race. During the year, no contact was permitted between Honeywell and Penn State. Each group needed to put in their maximum effort to win the competition next year. I learned that this development method (called the "NASA System") had accelerated space development in the US; leading US space technology — which had been far behind that of the former Soviet Union in the 1970s — to the later successful launch of the Space Shuttle.

Japanese "big projects" are either led by the Ministry of Economics, Trading and Industry (METI) or Ministry of Education, Culture, Science and Technology (MEXT). The project members are chosen from similar industrial group members in close communication with each other, leading to synchronous research (no competition). The NASA System seems to be advantageous from the view of increased creativity and research speed, but the Japanese style makes a good team with humanistic harmony [17]. Figure 4.9 depicts the differences in USA's and Japan's "big science project" structures. Note that there is only a vertical command line between the military program officer and the contracting researchers in the US system, while the horizontal interaction is strong among researchers of the same level in the Japanese system. The former structure

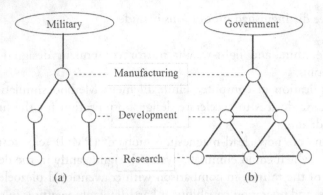

Fig. 4.9.   Difference in structure of "big science" projects, between (a) the US and (b) Japan.

seems to be suitable for long-term big science, while the latter structure fits short-term small developments.

According to the statistics for research and development investments in the US and Japan in 1980; in the United States, 50% of its R&D funds come from the military — federal institutes such as the Defense Advanced Research Project Agency (DARPA), Army, Navy, Air Force, etc. Among the 50% of R&D funds from non-military sources, a large portion came from non-military federal institutes such as NASA, National Institute of Health (NIH), National Science Foundation (NSF), etc., and the contribution by private industries is rather small. In addition, the military's share is significantly increasing. This investment situation accelerates the R&D structure initiated and controlled by the government, leading to the vertical command line shown in Fig. 4.9(a). In contrast, the Japanese government funds less than 30% of Japan's R&D, with a slight annual increase; the remaining more than 70% of the R&D is funded by private industries. Thus, though Japanese projects are initiated or flagged by the government, the projects are propelled by private industries. In such a situation, the friendly horizontal interaction shown in Fig. 4.9(b) becomes reasonable.

4.4 Summary

In this chapter, I presented how "Politico-Engineering" covers (1) crisis technologies, and (2) legally regulated normal technologies such as sustainability technologies. A paradigm shift is required for crisis/sustainability technology developments. We also discussed how crisis technologies can be further classified into technology for:

- Natural disasters (e.g., earthquakes, tsunamis, tornadoes, hurricanes, lightning).
- Epidemic/infectious diseases (e.g., smallpox, polio, measles, HIV)
- Enormous accidents (e.g., Three Mile Island core meltdown accident, BP oil spill)
- Intentional accidents (e.g., acts of terrorism, criminal activity)
- Civil wars, wars, and territorial aggression.

And that sustainability technologies include:

- Power and energy (e.g., solutions to the problem of a lack of oil, nuclear power plants, new energy harvesting)
- Rare material (e.g., rare-earth metal, Lithium)
- Food and Biofuel (e.g., rice, corn)
- Toxic materials

   o Restriction [of the use of] (e.g., heavy metals, Pb, Dioxin)
   o Elimination/neutralization [of] (e.g., mercury, asbestos)
   o [Search for] Replacement material

- Solutions for environmental pollution
- Technology to improve energy efficiency

We also learned that in politico-engineering, the legal regulation or a government's requirements for the final products come first, with strict constraints in terms of the date of issue of the law, and performance regulation (for example, the permitted ratio of Pb in the final product; the amount of CO_2, NO_x, or SO_x emissions; or the specifications for rescue robots and new weapons).

References

[1] P. Low, *Natural Catastrophes in 2012   Dominated by   U.S. Weather Extremes*, Worldwatch Institute, May 29, (2013).
[2] http://en.wikipedia.org/wiki/Ebola_virus_disease.
[3] http://en.wikipedia.org/wiki/Three_Mile_Island_accident.
[4] *14-Year Cleanup at Three Mile Island Concludes*, New York Times, August 15, (1993).
[5] http://en.wikipedia.org/wiki/Deepwater_Horizon_oil_spill.
[6] http://en.wikipedia.org/wiki/September_11_attacks.
[7] http://www.cbsnews.com/news/french-officials-deny-reports-of-germanwings-flight-9525-final-seconds-video/.
[8] http://theinstitute.ieee.org/ieee-roundup/opinions/ieee-roundup/protecting-critical- infrastructures-from-cyberattacks.

[9] http://en.wikipedia.org/wiki/Vietnam_War.

[10] http://en.wikipedia.org/wiki/Senkaku_Islands_dispute.

[11] http://en.wikipedia.org/wiki/Three_Gorges_Dam.

[12] R. Buerk, *Japan to free Chinese boat captain*, BBC September 24 (2010). Retrieved August 08, 2012.

[13] http://www.bbc.com/news/world-africa-21087732.

[14] http://en.wikipedia.org/wiki/Minamata_disease.

[15] http://www.rohsguide.com/rohs-substances.htm.

[16] http://en.wikipedia.org/wiki/Particulates.

[17] K. Uchino, *Entrepreneurship for Engineers*, CRC Press, Boca Raton, FL (2010), ISBN: 978-1-4398-0063-8.

CHAPTER 5

Crisis Technology

The world's largest earthquake and tsunami attacked the northern part of Japan's Honshu Island on March 11, 2011, killing 16,000 people. Furthermore, the problem of the accident at the Fukushima Daiichi Nuclear Power Plant that had been induced by the tsunami seemed to be more serious than the earthquake/tsunami incident itself — though no one had been directly killed by the Power Plant accident. The negative effects of the incident manifested over a long stretch of time rather than immediately, unlike the tsunami. In fact, five years after the accident, the Fukushima problem (i.e., Cs contamination) still persisted.

We consider crisis technologies from scientific and technological viewpoints in this chapter, by adopting practical examples; in particular, from our developments on piezoelectric actuators and sensors. As already introduced, there are five types of "crises":

1. Natural disasters
2. Epidemic/infectious diseases
3. Enormous accidents
4. Intentional accidents
5. Civil wars, wars, and territorial aggression

As infectious or contagious diseases can involve some association with terrorist activities, these five crises are sometimes related to each other. In the United States, politicians were attacked with anthrax in 2001, which was a combination of an epidemic and an intentional accident. *A disaster* is what happens in an accident and is not intentionally caused; while the term *crisis* contains all — whether they are intentional or accidental.

5.1 Natural Disaster

Research efforts into crisis engineering for natural disasters will be highly required. The research themes of urgent need in the area of actuators/sensors are the following:

(1) Accurate monitoring and surveillance techniques (e.g., vibration and pressure sensors) for earthquakes, tsunamis, typhoons, or tornados;
(2) Technologies for gathering and managing crisis information and informing the public (e.g., directive loudspeakers) in ways that do not evoke panic (this will be discussed in Chapter 7, "Crisis Management");
(3) Rescue technologies (autonomous unmanned underwater, aerial, surface/land vehicles, robots, etc.).

5.1.1 *Earthquake/Volcanic Eruption*

Earthquake/volcanic eruption prediction and anti-earthquake/preventive structure reinforcement technologies are discussed in the following segments.

5.1.1.1 *Earthquake Prediction Technologies*

When I was working in Japan, earthquake prediction research (often in the form of short-term projects; in the matter of months or days) was popular, as well as the development of rescue robots. However, because of the Japanese government's change in political strategy, they stopped funding crisis-related technologies over the last 20 years. Earthquake research in Japan during that period provided only long-term earthquake probability calculations, like this one: "The University of Tokyo's Earthquake Research Institute predicts there is a 70% probability that the Tokyo's metropolitan area will experience a magnitude-7 quake within four years and a 98% probability within the next 30 years" [Report by Science Council of Japan, Asahi Newspaper] — which neither conveys the message of urgency nor the necessity of preparation to the Japanese people; because the idea that "I will not be alive 30 years later ..." is a normal attitude for the average person. It was because of these — the combination of a governmental shift in policy, a decrease in funding for crisis technologies, and the lackadaisical attitude of humans — that resulted in unexpectedly huge damages when the Kanto–Tohoku Big Earthquake and Tsunami hit Japan on March 11, 2011.

In practice, there are primarily two methods for predicting earthquakes in the short-term: electromagnetic field monitoring and land displacement measurement.

Electromagnetic Field Monitoring: The earth's magnetic field is the result of large convection currents of magnetic iron within the earth's interior. It shields us from the intense blasts of solar radiation that would otherwise scorch our planet. However, recent studies have related the effects of the fluctuations in the earth's magnetic field to *seismic, and thermal events across the globe*, as well as several other inherent properties of organisms [1].

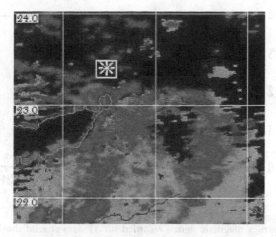

Fig. 5.1.   Infrared (IR) fluctuations between 6°C and 9°C days before a 1998 Zhangbei earthquake.

Friedemann Freund (San Jose State University, USA) referenced Russian and Chinese Scientists analyzing Infrared (IR) data in the 1998 Zhangbei earthquake near the Great Wall of China. In Fig. 5.1, IR fluctuations between 6°C and 9°C were noted days before an earthquake. Interestingly, research indicated that the *IR energy* released was neither resultant from tectonic friction nor magma proximity, but electrical voltage buildup [2]. Freund *et al.* placed a granite rock under a 1,500 ton press and demonstrated IR and electrical emissions. The slab might react similarly to a semiconductor, generating and combining electron–hole pairs (or positive ions), which would release IR energy in the process.

Electrical currents in rocks might explain another curious observation: scientists doing research with magnetometers just before major earthquakes have serendipitously recorded tiny, slow fluctuations in Earth's magnetic field. One example happened during the Loma-Prieta earthquake that devastated San Francisco in 1989. Almost two weeks before the quake, readings of *low-frequency magnetic signals* (0.01–0.02 Hz) jumped up to 20 times above normal levels, and then spiked even higher on the day of the quake (see Fig. 5.2) [2].

My group at Penn State University invented magenetoelectric composite sensors by laminating piezoelectric and magnetostrictive plates, which demonstrated extremely effective high resolution and sensitivity for low- frequency low-amplitude magnetic fields [3]. Figure 5.3 shows a prototype structure of the device, in which a PZT disk is sandwiched by two Terfenol-D (magnetostrictor) disks. When a magnetic field is applied on this composite, Terfenol-D expands; the reaction is mechanically transferred to PZT, leading to the generation of an

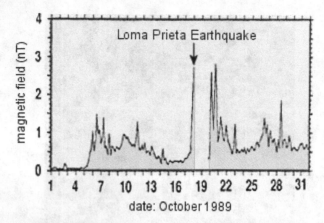

Fig. 5.2.   Low-frequency magnetic signals recorded for 31 days around the 1989 Loma-Prieta earthquake.

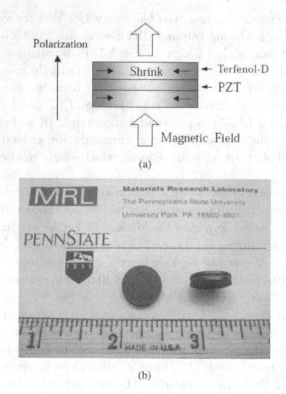

Fig. 5.3.   Magnetic noise sensor consisting of a laminated composite of a PZT and two Terfenol-D disks. (a) Schematic principle and (b) sensor picture.

electric charge from the PZT. By monitoring the voltage generated in the PZT, we can detect the magnetic field. The key of this device is high sensitivity at a low frequency.

Extremely low equivalent magnetic noise in a Metglas/piezofiber magnetoe-lectric (ME) magnetic-field sensor, realized through a combination of a giant ME effect and a reduction in constituent internal noise sources, was demonstrated by a Virginia Tech group [4]. The ME coefficient is 52 V cm^{-1} Oe^{-1} at low frequency; the 1 Hz equivalent magnetic noise is 5.1 pT Hz$^{-1/2}$; and the magnetic field sensitivity is 10 nT at a significantly lower cost than current devices (SQUID) can achieve.

Land Displacement Monitoring: The earth's crust also deforms in the months and weeks leading to a seismic event. NASA has tracked these fluctuations, and formed dynamic contour maps at a resolution of 1mm/year to map the changes in Earth's crust [2].

An ultra-high resolution geological surveying algorithm that splices data, creating a dynamic histogram of Earth's crust's changing altitude to a resolution of 1 mm per year, can display a likelihood of an earthquake based on how much land has change height in how long of a timeframe. There are several satellite-based methods that show promise as precursors to earthquake activity, including Interferometric-Synthetic Aperture Radar (*InSAR*). Basically, InSAR is when two radar images of a given tectonic area are combined in a process called data fusion, and any changes in ground motion at the surface may be detected. Figure 5.4 shows an InSAR image showing the shift in ground height due to the 1999 Hector Mine earthquake. The radar data were acquired by the European Space

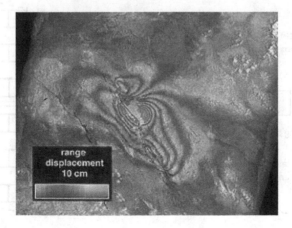

Fig. 5.4.   An InSAR image showing the shift in the ground height due to the 1999 Hector Mine earthquake.

Agency's ERS-2 satellite on September 15 and October 20, 1999 (before and after the earthquake).

5.1.1.2 *Anti-Earthquake Technologies*

Pre-Stressed Structures: Are the intentional creation of permanent stresses in a structure for the purpose of improving its performance under various service conditions. Pre-stressed structures are widely used in buildings, bridges, underground structures, and nuclear reactor vessels.

Reinforced Concrete Structures: Reinforced concrete is concrete in which steel reinforcement bars (*rebar*) have been incorporated to strengthen a material that would otherwise be brittle. Pre-stressed concrete is used to overcome concrete's natural weakness in tension. This is done by pre-stressing rebar in tension while the concrete is poured. Once the concrete hardens, the rebar induces a compressive force on the concrete. To combat failure due to earth shaking, concrete structures also have a number of ductile joints to allow structural mobility.

The *super-elastic behavior* of *shape memory alloys* (*SMAs*) has attracted the attention of civil engineers [5]. The Nickel–Titanium alloy (NiTiNOL) wire delivers a constant force over an extended portion of the strain range (up to 5%). Its major field of application is retrofitting structures in an earthquake design. A real scale application of a superelastic SMA device is the earthquake-resistant retrofit of the Basilica San Francesco at Assisi, Italy (*M. G. Castellano*). The historic gable was connected to the main structure by devices using SMA rods, as shown in Fig. 5.5. The Ni–Ti SMA rods were subjected to tension although

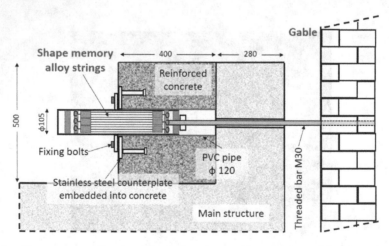

Fig. 5.5.   Shape memory alloy device for earthquake-suitable connection of the historic gable and the main structure of the Basilica San Francesco in Assisi, Italy.

they were designed to take tension and compression forces. Regardless of the structure deformation, the SMA alloy keeps the same compressive force to the concrete and the gable.

Active/Passive Damping: "Lead rubber bearing" is a base isolation employing heavy damping. A *heavy damping mechanism* is incorporated in base isolation devices to suppress vibration, thus enhancing a building's seismic performance. *Tuned mass dampers* are used in skyscrapers or other structures and move in opposition to the resonance frequency in order to reduce the lateral movement of the building. A friction pendulum bearing allows a building to be isolated from its foundation. Its purpose is to minimize damage caused by lateral displacements observed during earthquakes. The *building elevation configuration* in a pyramid- shaped building offers better stability. The earthquake or wind quieting ability of the elevation configuration is provided by a specific pattern of multiple reflections and transmissions of vertically propagating waves, which are generated by break- downs into homogeneity of story layers, and a taper.

I lived in Roppongi Hills, an apartment house developed by Mori Building, during my stint as a Navy Officer. The building installs 192 hydraulic oil dampers (shock absorbers) and braces made of flexible steel, which were designed to minimize sway caused by earthquakes or heavy winds. Semi-active oil dampers control tremor by the flow of their oil. The other type of damper used is the Unbonded Brace, which consumes yielding energy through its pliant, yet durable structural steel covered with concrete. These are positioned all the way up the 238 m building. A *Concrete Filled Steel Tube* (CFT) consists of a steel tube filled with concrete. A pillar made of CFT structure complements the weaknesses of each material, attaining higher strength against buckling by earthquake or heat, as shown in Fig. 5.6 [6].

Fig. 5.6.   Anti-seismic Concrete Filled Steel Tube structure in the Roppongi Hills buildings.

5.1.2 *Tsunami*

5.1.2.1 *Tsunami Prediction Technology*

Fujitsu Ltd., Japan, in collaboration with Tohoku University, Japan, developed a 3D tsunami simulator which can replicate the surge of water in urban areas as well as river surges caused by a tsunami in fine detail [7]. The joint research combined a 2D tsunami-propagation simulation technology developed by *Fumihiko Imamura* (Tohoku University) with Fujitsu's 3D fluid simulation technology. As a result, the researchers were able to accurately replicate the complex changes a tsunami undergoes as it interacts with coastal topography or buildings in urban areas, as well as the process of water surges in urban areas and rivers, as shown in Fig. 5.7.

5.1.2.2 *Tsunami Protection Technology*

Japan is planning to build a 400 km chain of tsunami protection sea walls to fend off any future natural disasters in the Tohoku area. Four hundred and forty sea walls will be built, aimed at stopping a repeat of the devastating 2011 tsunami, which crippled the country and led to the Fukushima nuclear disaster. Some parts of the $6.8 billion project will reach a height of five stories, which earned it the nickname *The Great Wall of Japan* (see Fig. 5.8). But this is not final, since earlier reports from the ministries of agriculture and land said that it would require the building of walls on 14,000 km of Japan's 35,000 km coastline.

Opponents of the massive structure say it is a waste of money, most of which will be hived off from regional budgets. Traditional sceneries will be disrupted,

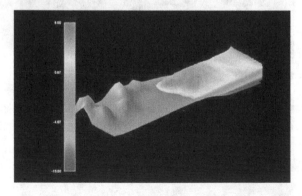

Fig. 5.7.   Replication of the behavior of a tsunami as it flows into coastal areas near Sendai, Japan, in 3D.

Fig. 5.8. Tsunami protection "The Great Wall of Japan" in northern Japan, Tohoku area.

and it could damage marine life; while the risk zone areas will not be totally disaster-proof. Sea walls are a controversial issue. *Fudai,* a village sheltering behind a giant concrete shield, escaped intact in 2011. However, 90% of existing seawalls along the northeast coast were not as successful: a $1.6-billion breakwater protecting the city of *Kamaishi* crumbled on impact, and some 1,000 people died there.

5.1.3 *Tornado*

A *tornado* (or twister) is a violently rotating column of air that is in contact with both the earth's surface and a cumulonimbus cloud. Tornadoes come in many shapes and sizes, but they typically form a visible condensation funnel whose narrow end touches the earth, and are often encircled by a cloud of debris and dust. Most tornadoes have wind speeds less than 180 km/h, are about 80 "macros," and travel several kilometers before dissipating.

Tornadoes have been observed on every continent except Antarctica. However, the vast majority of tornadoes occur in *Tornado Alley* in the United States, although they can occur nearly anywhere in North America. They also occasionally occur in South-central and Eastern Asia, Northern and East-central South America, Southern Africa, Northwestern and Southeast Europe, Western and Southeastern Australia, and New Zealand. Tornadoes can be detected before or as they occur through the use of *Pulse-Doppler radar* by recognizing patterns in velocity and reflectivity data (such as hook echoes or debris balls), as well as through the efforts of storm spotters. There are several scales for rating the strength of tornadoes. The *Fujita scale* rates tornadoes by damage caused and has been replaced in some countries by the updated Enhanced Fujita Scale. An F0 or EF0 tornado, the weakest category, damages trees, but not substantial structures. An F5 or EF5 tornado, the strongest category, rips buildings off their foundations and can deform large skyscrapers. Doppler radar data, photogrammetry, and ground swirl patterns (cycloidal marks) may also be analyzed to determine intensity and assign a rating [8].

Meteorology is a relatively young science, and the study of tornadoes is newer still. Scientists have a fairly good understanding of the development of thunderstorms and mesocyclones, and the meteorological conditions conducive to their formation; however, the step from supercell (or other respective formative processes) to *tornadogenesis* and predicting tornadic vs. non-tornadic mesocyclones is not yet well known and is the focus of much research.

5.1.4 *Hurricane/Typhoon/Flood*

Hurricanes, cyclones, and typhoons are all the same weather phenomenon; we use different names for these storms in different places. In the Atlantic and Northeast Pacific, the term *hurricane* is used. The same type of disturbance in the Northwest Pacific is called a *typhoon*, and *cyclones* occur in the South Pacific and Indian Ocean.

The ingredients for these storms include a pre-existing weather disturbance, warm tropical oceans, moisture, and relatively light winds. If the right conditions persist long enough, they can combine to produce the violent winds, incredible waves, torrential rains, and floods associated with this phenomenon. In the Northwest Pacific, typhoon season officially runs from June 1 to November 30. However, while 97% of tropical activity occurs during this time period, there is nothing magical in these dates, and hurricanes have occurred outside of these six months.

Hurricanes start when warm, moist air from the ocean surface begin to rise rapidly. At the point where the warm air encounters cooler air, the warm water vapor condense and form storm clouds and drops of rain. The condensation also releases latent heat, which warms the cool air above, causing it to rise and make way for more warm humid air to rise from the ocean below. As this cycle continues, warmer moist air is drawn into the developing storm and more heat is transferred from the surface of the ocean to the atmosphere. This continuing heat exchange creates a wind pattern that spirals around a relatively calm center, or *eye*, like water swirling down a drain. Since a *microcyclone* is initially created on a tropical ocean, we can usually have a sufficient time period (at least 3–4 days) to predict which residential areas the hurricane will hit, and prepare for the evacuation of the inhabitants.

5.1.5 *Lightning*

As already introduced in Chapter 1, *Benjamin Franklin* used a child's kite to prove that lightning is really a stream of electrified air. Lightning is a sudden electrostatic discharge during an electric storm between electrically charged regions of a cloud (called intra-cloud lightning or IC), between that cloud and another cloud (cloud to cloud, or CC, lightning), or between a cloud and the ground (CG lightning). The charged regions within the atmosphere temporarily equalize themselves through a lightning flash — commonly referred to as a strike if it hits an object on the ground. Although lightning is always accompanied by the sound of thunder, distant lightning may be seen but may be too far away for the thunder to be heard.

Lightning primarily occurs when warm air is mixed with colder air masses, resulting in atmospheric disturbances necessary for polarizing the atmosphere. About 90% of *ionic channel* lengths between "pools" are approximately 45 m in length. The establishment of the ionic channel takes hundreds of milliseconds, in comparison to the resulting discharge, which occurs within a few milliseconds. The electric current needed to establish the channel, measured in the tens or hundreds of A, is dwarfed by subsequent currents during the actual discharge. The most effective method to protect residents is to use a *lightning rod* (or lightning protector), which is a metal strip or rod connected to earth through conductors and a grounding system; to provide a preferred pathway to ground if lightning terminates on a structure. A lightning rod, also known as a "Franklin rod" in honor of its inventor, *Benjamin Franklin*, is simply a metal rod when not connected to the lightning protection system (i.e., *arresters*). Modern lighting arresters, constructed with metal oxides, are capable of safely shunting abnormally high voltage surges to ground while preventing normal system voltages from being shorted to ground.

Commonly used is the *metal-oxide varistor* (MOV), which contains a ceramic mass of *ZnO* grains, in a matrix of other metal oxides (such as small amounts of bismuth, cobalt, and manganese) sandwiched between two metal plates (i.e., electrodes) (see Fig. 5.9). The boundary between each grain and its neighbor forms a *diode junction*, which allows current to flow in only one direction. The mass of randomly oriented grains is electrically equivalent to a network of back-to-back diode pairs, with each pair in parallel with many other pairs. When a small or moderate voltage is applied across the electrodes, only a tiny current flows, caused by reverse leakage through the diode junctions. When a large voltage is applied, the diode junction breaks down due to a *combination of thermionic emission and electron tunneling*, and a large current flows. The result of this behavior is a highly nonlinear current–voltage characteristic, in which the MOV has a high resistance at low voltages and a low resistance at high voltages [9].

Fig. 5.9.   A high voltage MOV.

5.2 Epidemic/Infectious Disease

According to the World Health Organization (WHO), *infectious diseases* are defined as diseases caused by pathogenic microorganisms (*germs*), such as *bacteria*, *viruses*, *parasites*, or *fungi*; the diseases can be spread, directly or indirectly, from one person to another. *Zoonotic diseases* are infectious diseases in animals that can cause illnesses in humans when transmitted from animals to humans. Swine flu (swine influenza), a respiratory disease of pigs caused by viruses, is a recent example of this. In a number of instances, people were infected by swine flu from being closely associated with pigs (for example, pig farmers and pork processors); and likewise, pig populations have occasionally been infected with the human flu. Swine flu is transmitted from person to person by the inhalation or ingestion of droplets containing the virus, from people sneezing or coughing; it is not transmitted by eating cooked pork products [10]. *Bacteria* are single-celled microorganisms that exist in abundance in both living hosts and in all areas of the planet (e.g., soil, water, etc.), while a *virus* is acellular (has no cell structure) and requires a living host to survive; it causes an illness in its host, which then elicits immune responses.

There are multiple phase crisis technologies for epidemics/infectious diseases:

(1) Detecting/monitoring infectious/contagious disease germs,
(2) Disinfecting the germs,
(3) Curing patients via medicine or vaccines.

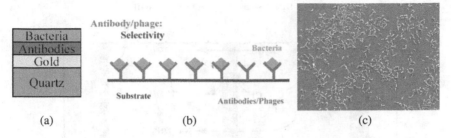

Fig. 5.10. Bacteria sensor: (a) basic structure with quartz membrane, (b) principle of bacteria selective capturing by antibodies/phages, and (c) scanning-electron-microscope's picture of captured bacteria. [Courtesy of Z.-Y. Cheng at Auburn University, USA.]

5.2.1 *Detection of Infectious Disease Germs*

Quartz membrane is occasionally used for various micro-mass sensors. Because the mechanical quality factor Q_m is very large (~10^6), the monitoring resolution of the resonance frequency reaches $\Delta f_R/f_R \sim 10^{-6}$. Thus, even small mass changes on the quartz crystal surface can be finely detected through the resonance frequency shift. This micro-mass sensor can be utilized as a bio-sensor for detecting bacteria. *Z.-Y. Cheng* at Auburn University, USA, reported the results on *E. Coli and Salmonella* via private communication in 2003. Figure 5.10 illustrates the bacteria sensor's basic structure with quartz membrane, the principle of bacteria selective capturing by *antibodies/phages*, and the scanning electron microscope's picture of captured bacteria. A specific antibody/phage can be coated on a single crystal quartz oscillator. Once particular bacteria are captured selectively by the antibodies, the surface mass of the oscillator is increased. A sensitivity of 10^4 cells per mL has been reported, which is sufficient to monitor 10^4–10^7 cells per mL — a quantity fatal to humans in the case of Salmonella.

5.2.2 *Disinfection of Disease Germs*

The 2001 *anthrax* attacks in the US, also known by its Federal Bureau of Investigation (FBI) case name, *Amerithrax*, occurred over the course of several weeks beginning on Tuesday, September 18, 2001 — one week after the September 11 attacks. Letters containing anthrax bacteria spores were mailed to several news media offices and two Democratic US Senators, killing five people and infecting 17 others. According to the FBI, the ensuing investigation became "one of the largest and most complex in the history of law enforcement."

In order to neutralize the biological weapon, my group at Penn State University developed a portable *hypochlorous acid* (HClO) disinfection device using a piezoelectric ultrasonic humidifier [11]. Hypochlorous acid has proven to be a strong disinfectant with no side effects on humans after a couple of hours

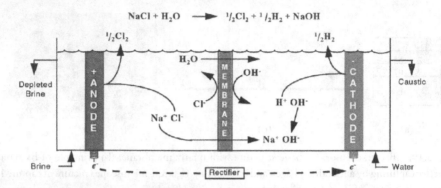

Fig. 5.11.   Generation of chlorine and caustic in a membrane electrolytic cell.

of its preparation, and would be ideal against *bacteria* like Anthrax, and for disinfecting offices and hospital buildings. Coupled with the atomization of the acidic solution, much higher disinfection effects can be expected within a suitable time period. The acid is not sold as a pure solution, since it naturally disintegrates after a few hours. A corrosion-resistant electrolytic cell was designed to produce $HClO$ from brine. Nafion® membranes were used to separate the anode and cathode compartments and produce $HClO$ and sodium hydroxide in the anode and cathode compartments, respectively. An ultrasonic piezoelectric atomizer was utilized to generate micro-droplets of the diluted acid. Figure 5.11 shows the general configuration of the electrolytic cell. The membrane in the cell is permeable to many cations and polar compounds but can almost completely reject anions and nonpolar species. Therefore, sodium ions will be trapped in the cathode compartment preventing them from reacting with the ClO^- ions to produce $NaClO$, which is not as strong a disinfectant as $HClO$.

5.2.3 *Vaccine/Medicine*

The 2014 *Ebola* epidemic is the largest in history. Affected States in West Africa included Guinea, Liberia, and Sierra Leone. Japanese Fujifilm Holdings Corporation's drug "Avigan," which has shown signs of efficacy against the Ebola virus, has drawn interest from about 20 countries and the company stands ready for large orders [12]. Avigan was developed by a Fujifilm subsidiary, Toyama Chemical Co., and approved by Japanese regulators in March 2014 as an anti-influenza medicine. It subsequently emerged as a potential treatment for Ebola, which has killed more than 9,600 people during a recent outbreak centered in West Africa. The French National Institute of Health and Medical Research released preliminary results of a trial in Guinea, suggesting that the drug could cut the number of deaths among Ebola patients with moderate to high levels of the virus by half. Those suffering from higher levels of infection did not appear to benefit from the drug.

On the other hand, researchers at Thomas Jefferson University in Philadelphia, USA, are spearheading another research effort along similar lines. They have produced an Ebola virus vaccine that piggybacks on the established rabies virus vaccine, and the new combination has successfully immunized mice and primates in lab tests against both rabies and Ebola. After these animal tests, they have to produce a vaccine that would be appropriate for humans [13].

5.3 Enormous Accident

The Three Mile Island accident was a partial nuclear meltdown that occurred in 1979 in one of the two Three Mile Island nuclear reactors in Pennsylvania, USA [14]. It was the worst accident in the history of US commercial nuclear power plants. The incident was rated "5" on the 7-point International Nuclear Event Scale; i.e., an "Accident with Wider Consequences."

I was a Research Associate at the Materials Research Laboratory (MRL), Penn State University, when this meltdown occurred. The accident began at about 4 am on Wednesday, March 28, 1979. However, new concerns arose by the morning of Friday, March 30. A significant release of radiation from the plant's auxiliary building, performed to relieve pressure on the primary system and to avoid curtailing the flow of coolant to the core, caused a great deal of confusion and consternation. In an atmosphere of growing uncertainty about the condition of the plant, the then-governor of Pennsylvania, *Richard L. Thornburgh*, consulted with the Nuclear Regulatory Commission (NRC) about evacuating the population within an 8 km radius of the plant. Since our State College township was outside of the evacuation area, we did not feel any panic. Then-Director of MRL, *Rustum Roy*, called all MRL researchers to work on the rescue research programs two weeks after this accident. Though the area was not my area of expertise, I was involved in the project developing new cement to prevent/minimize the contamination of the environment around the plant. In central Pennsylvania (where we were), extensive underground *"rock salt"* beds extend from the Appalachian basin of western New York through parts of Ontario. Roy pointed out that "Since regular cement cannot solidify under salty conditions, we will need a new cement which can quickly solidify even under sea water; that can completely seal-in the affected nuclear reactor." It took about half a year to develop the cement.

As described multiple times previously, the Fukushima Daiichi nuclear power plant meltdown occurred as a result of the earthquake/tsunami, followed

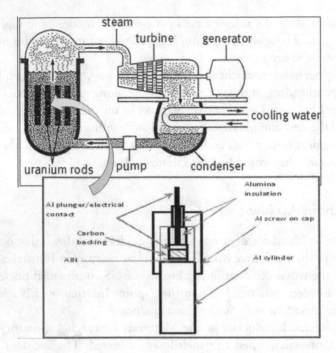

Fig. 5.12. Aluminum nitride piezoelectric transducer for monitoring the uranium rod condition in a nuclear chamber. [Courtesy of B. R. Tittmann, Penn State University.]

by an explosion. Because the explosion happened owing to a lack of systems for monitoring the meltdown (Japanese nuclear power plants did not have even the most primitive sensing system, to tell the truth!), *B. R. Tittmann et al.* at Penn State are currently developing a high-temperature (600°C) operating piezoelectric transducer with an aluminum nitride (AlN) single crystal for monitoring the condition of uranium rods in a nuclear chamber, as shown in Fig. 5.12 [15]. Using a piezoelectric transducer, it is theoretically relatively easy to monitor status changes in uranium rods. But, there are two key issues to practical developments: (1) limitations in the high-temperature reliability of the piezoelectric material (most commercially available materials cannot be used in temperature ranges higher than 250°C), and (2) limited knowledge in high radiation durability (we do not have sufficient durability test data under numerous radiation conditions). Though studies on the effects of radiation on piezoelectric/ferroelectric materials were rather popular in the 1970s in the former Japan Atomic Energy Research Institute (JAERI), the Japanese government reorganized the laboratories and discouraged research on this topic after the 2000s; merely because it might create a bad reputation for the Nuclear Power Plant project among the Japanese people by suggesting that a nuclear power plant might cause an accident. However, this governmental misjudgment on the development of safety technology and even on diminishing existing security technologies (such as rescue robots) have actually enhanced the extent of the disaster's damage in Tohoku area. This Fukushima incident is discussed again in Chapter 7, (Risk Management).

5.4 Intentional Accident

In this section, we discuss acts of violence or attacks by non-state actors (individuals or groups).

5.4.1 *Acts of Terrorism, Criminal Activities*

5.4.1.1 *Six Types of Terrorism*

Terrorism is an act of politically, ideologically or religiously motivated *violence by non-state actors*. According to the National Advisory Committee on Criminal Justice Standards and Goals, there are six distinct types of terrorism [16]. All of them share the common traits of being violent acts that destroy property, invoke fear, and attempt to harm the lives of civilians.

1. *Civil disorder* is violence in the form of a large riot or protest held by a group of individuals; usually in opposition to a political policy or action. Such protests are intended to send a message to a political group that "the people" are unhappy and demand change.
2. *Political terrorism* is used by one political faction to intimidate another. Although government leaders are the intended recipients of the ultimate message, it is the citizens who are targeted with violent attacks.
3. *Non-political terrorism* is a terrorist act perpetrated by a group for any other purpose, most often of a *religious nature*. The desired goal is something other than a political objective, but the tactics involved are the same.
4. *Quasi-terrorism* is a violent act that utilizes the same methods terrorists employ, but does not have the same motivating factors. The *law breaker* is acting in a similar manner to a terrorist, but terrorism is not the goal.
5. *Limited political* acts of *terrorism* are generally one-time only plots to make a political or ideological statement. The goal is not to overthrow the government, but to protest a governmental policy or action.
6. *State terrorism* is a violent act initiated by an existing government to achieve a particular goal. Most often, this goal involves a conflict with another country.

These terroristic attacks may occur at any time or place, which makes them an extremely effective method of instilling terror and uncertainty into the general public.

5.4.1.2 *Anti-Terrorism Technologies*

Most States' anti-terrorist policies and measures include the development and/or adoption of anti-terrorism technologies. These terrorist surveillance technologies include visual, acoustic, x-ray methods, etc.

Security Cameras: These are one of the most popular visual surveillance gadgets and can be installed anywhere, including in banks, at public places, along roads, as well as in/around individual homes. The visual data captured is matched (e.g., through face and fingerprint recognition software) with already available *biometric data* of terrorists to prevent criminal attacks. As shown in Fig. 5.13, face and fingerprint recognition software have become popular from an information security viewpoint even for laptop computers and room entrance systems. In fact, technology that can be used for such purposes has made putting a photo of yourself with a 'peace' (or 'V') sign on your Facebook page dangerous, because the resolution of the photos taken by smart phone cameras is high enough for hackers to obtain your fingerprint and replicate it — at a level sufficient to get past biometric security measures.

Acoustic Detection Systems: The utilization of acoustic technology to detect terrorists hidden in ground vehicles at border checkpoints between two states/countries and in cargo containers at ports, are currently under consideration. The heartbeat of terrorists hidden in ground vehicles (e.g., trucks, busses, etc.)

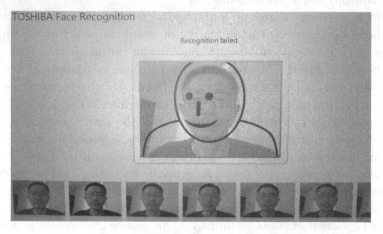

Fig. 5.13.   Face recognition function on a laptop computer.

can be detected by and monitored with sensitive piezoelectric acoustic sensors embedded on the ground — even via the vehicle tires. Such a system can also identify the number of terrorists and even their sexes (male (or) female), since the cyclical heartbeat is a sort of biometric ID for humans.

Airport Security Systems: As you are aware, in addition to the *metal detectors* meant to detect weapons like guns and knives, most airports are equipped with *x-ray* machines for the purpose of monitoring the transportation of illegal material (e.g., explosives, drugs, etc.) via airplanes.

Gamma-Ray Machines: You may feel frustrated when airport security removes what is no more than a bottle of water from your carry-on baggage at the checkpoint every time. This is because the X-ray system cannot distinguish water from liquid bombs, such as NTN (nitroglycerin and triacetonetriperoxide). My group at Micromechatronics Inc., PA (a spin-off from Penn State University), in collaboration with Lawrence Livermore National Laboratory (LLNL), CA, is developing a *gamma-ray scan inspection system*, which can distinguish between atomic species, i.e., distinguish liquid explosives from water. Gamma rays are the highest-energy form of electromagnetic radiation, being physically the same as all other forms (e.g., X-rays, visible light, infrared, radio) but having higher photon energy due to their shorter wavelength. Because of this, the energy of gamma-ray photons can be resolved individually, and a gamma-ray spectrometer can measure and display the energies of the gamma-ray photons detected. LLNL developed a compact neutron accelerator, as pictured in Fig. 5.14(b). The key to this system is miniaturizing the 1 MV voltage supply

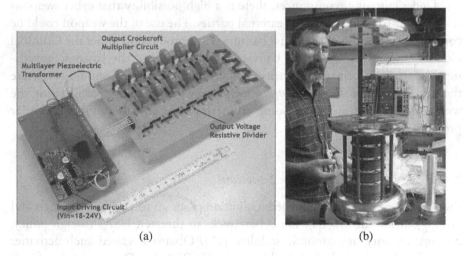

(a)   (b)

Fig. 5.14. (a) 100 kV to 1 MV voltage source with piezoelectric transformers and (b) compact neutron accelerator for generating gamma-rays.

to accelerate a neutron beam. We utilized multilayer piezoelectric transformers, combined with Cockcroft multiplier circuits (Fig. 5.14(a)).

5.4.2 Cyber Threats

5.4.2.1 Cyber Weapon

In 2012, Fujitsu Ltd., Japan, developed a *cyber weapon* (i.e., computer virus) capable of tracking, identifying, and disabling sources of cyberattacks [17]. Now, cyber weapons are already in use in countries such as the United States and China. However, since there is no provision in the existing legislation on the use of cyber-weapons against external parties that wage attacks in Japan, this weapon has been tested in a closed network environment.

The Defense Ministry's Technical Research and Development Institute, which is in charge of weapons development, outsourced this project to Fujitsu Ltd. The most distinctive feature of the new virus is its ability to trace the sources of cyberattacks. It can identify not only the immediate source of attack, but also the "springboard" computers that were used to transmit the virus. The virus also has the ability to disable the attacking program and collect relevant information. According to the sources, the program can identify the source of a cyberattack to a high degree of accuracy for distributed denial of service (DDoS) attacks, as well as some attacks aimed at stealing information stored in target computers. In DDoS attacks, hackers send enormous volumes of data to target websites, eventually forcing them to shut down.

Under current circumstances, there is a high possibility that cyber weapons cannot be used in Japan against external parties. The use of the weapon could be considered a violation of the clause banning virus production under the Criminal Code. *Motohiro Tsuchiya* (Keio University), a member of the Japanese government panel on information security policy, opined that Japan should accelerate the development of *anti-cyber-attack weapons* by immediately reconsidering the weapon's legal definition, as some countries have already launched similar projects.

5.4.2.2 The US Cyber Attack Sanctions

President *Barack Obama* launched a sanction program to target individuals and groups outside the US that use cyberattacks to threaten USA's foreign policy, national security or economic stability [18]. Obama declared such activities as "national emergencies" and allowed the US Treasury Department to freeze assets and bar other financial transactions of entities engaged in destructive cyberattacks. The executive order gave the administration these sanction tools for deployment to also address other threats including the crises in the Middle East and Russia's aggression in Ukraine. These tools are now available against a growing epidemic of cyber threats aimed at destroying US computer networks. The effort to toughen the response to hacking follows indictments of five Chinese military officers and the decision to "name and shame" North Korea for a high-profile attack on Sony Pictures Entertainment in 2014. China routinely denies accusations by US investigators that hackers backed by the Chinese government have been behind the attacks on US companies. The said cyberattacks were generally cross-border incidents with hard-to-track origins.

Obama said:

> *"Harming critical infrastructure, misappropriating funds, using trade secrets for competitive advantage and disrupting computer networks would trigger the penalties. Companies that knowingly use stolen trade secrets to undermine the U.S. economy would also be targeted. From now on, we have the power to freeze their assets, make it harder for them to do business with U.S. companies, and limit their ability to profit from their misdeeds."*

The program was designed as a deterrent and punishment, filling a gap in US cybersecurity efforts where diplomatic means or means of law enforcement were insufficient.

However, the executive order was broad, which could result in a compliance nightmare for companies, and pundits warned that it remained difficult to definitively "attribute" hacking attacks and identify those responsible.

5.5 Civil War, War, and Territorial Aggression

5.5.1 *"Green" Weapons — Environment-Friendly Weapons*

Until the 1960s, the development of *weapons of mass destruction* (WMDs), including nuclear bombs and chemical weapons, was the primary focus of military-driven research. However, based on the global trend for *"Jus in Bello* (Justice in War)," *environmentally-friendly "green" weapons* (i.e., minimally destructive weapons with pin-point accuracy, such as laser guns and rail guns) became the mainstream focus in the 21st century.

The USS Ponce conducted tests for the Office of Naval Research-funded *Laser Weapons System* (LaWS) during the Persian Gulf in 2014. Directed energy weapons can counter asymmetric threats, including unmanned and light aircraft and small attack boats [19]. The tests were conducted under 20 kW using a 1 μm wavelength laser (see Fig. 5.15).

Fig. 5.15.   US Navy's Laser Weapons System.

Fig. 5.16.   US Navy's gigantic railgun.

A *rail gun* is an electrically powered electromagnetic projectile launcher, which comprises a pair of parallel conducting rails, along which a sliding armature is accelerated by the electromagnetic effect of a current that flows down one rail into the armature and then back along the other rail. Do you still remember the famous "right-hand rule" of Lorentz force for a current-carrying wire in a magnetic field B, learned in high-school physics? The *Hyper Velocity Projectile* (HVP) will be compatible with current weapons systems like the Navy 5-Inch Mk 45, and Navy, Marine Corps, and Army 155-mm Tube Artillery systems [20]. It was designed to be a guided projectile with low drag for high velocity and maneuverability, and decreased time-to-target. The HVP is 24 inches long and weighs 28 pounds (Fig. 5.16), and its ammunition is easy to handle and transport.

Fig. 5.17. Programmable air-bust munition (PABM) developed by ATK and MMech (25 mm caliber).

In this direction, *programmable air-burst munitions* (PABM), which are widely used in the battlefield in the Middle East, were successfully developed in 2004. The 25 mm caliber "Programmable Ammunition" by ATK Integrated Weapon Systems, Arizona, and Micromechatronics, Pennsylvania [21, 22], is illustrated in Fig. 5.17. Instead of an originally installed battery, a multilayer piezo-generator is used for generating electric energy under shot impacts (i.e., piezoelectric energy harvesting) to activate the operational amplifiers which ignite the burst according to the command program.

5.5.2 *Warfare Gadgets*

5.5.2.1 *Satellite Monitoring System*

The use of a precise wave front control (shown in Fig. 4.8), a deformable mirror with a monolithic piezoelectric, bulk ceramic to correct the distortion that occurs in a satellite telescope image due to atmospheric conditions (as illustrated in Fig. 5.18), in order to monitor the former Soviet Union's military bases, was proposed. The resolution requirement for the project (as I remember it) included a specification that required the system to be able to "distinguish a male from a female from a satellite!" However, the monolithic piezoelectric deformable mirror requires many electrodes with electrical leads to individual electrode elements; the total weight and volume (in particular, high voltage lead cables)

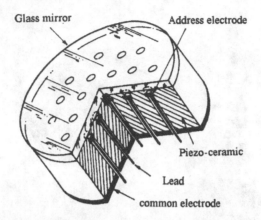

Fig. 5.18.   A deformable mirror made from a piezoelectric monolith.

were a serious hindrance to installing the mirror on a satellite. Thus, my Penn State group proposed a much simpler *multimorph deformable mirror*, which can be more simply controlled by means of a microcomputer [23]. As examples, let us consider the cases where we wish to refocus to correct for the coma aberration. The contour of the mirror surface can be represented by the *Zernike aberration polynomials*, whereby an arbitrary surface contour modulation $g(x, y)$ can be described as follows:

$$g(x, y) = C_r(x^2 + y^2) + C_c^1 x(x^2 + y^2) + C_c^{2\ 2} y(x^2 + y^2) + +$$

Notice that the Zernike polynomials are orthogonal and therefore completely independent of each other. The C_r and C_c terms represent *refocusing* and the *coma aberration*, respectively. As far as human vision is concerned, correction for aberrations, up to the second order in this series (representing astigmatism), is typically sought to provide an acceptably clear image. A uniform large area electrode can provide a parabolic (or spherical) deformation with the desired focal length. In the case of coma correction, one could employ the electrode pattern shown in Fig. 5.19(a), which consists of only six elements addressed by voltages applied in a fixed ratio. I was one of the world's first to apply the finite element computer simulation method on piezoelectric devices to accelerate design optimization. I remember that the calculation took half a day at the cost of US$20K just to optimize only six-segmented electrode patterns (10×10 matrix) and six levels of voltage on each electrode, in 1980.

Measurement of the deflection of a deformable mirror was carried out using a holographic interferometric system. Experimental results characterizing mirror deformations established for the purpose of refocusing, coma correction, and both refocusing and coma correction are shown in Fig. 5.19(b), which demonstrates

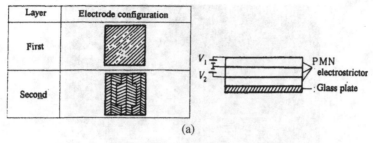

(a)

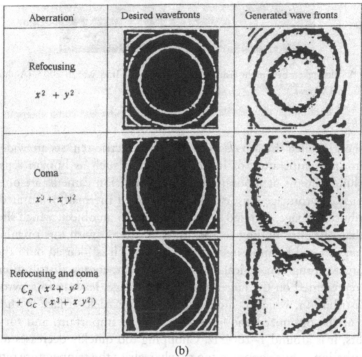

(b)

Fig. 5.19. A PMN-based multilayer deformable mirror: (a) diagrams of internal electrode configurations and the overall multilayer structure and (b) interferograms revealing the surface contours produced on the mirror.

the effectiveness of achieving a superposition of deformations by means of an appropriate configuration of discrete electrodes.

5.5.2.2 *Night Vision*

Ferroelectric materials exhibit the temperature dependence of spontaneous polarization, i.e., an electric field generated by changing the temperature. This

Fig. 5.20.   A video taken by a night vision camera during the Iraq War in 2003. (A photo taken by Uchino from TV news.)

phenomenon is called the *pyroelectric effect*. Pyroelectric sensors are widely used for monitoring temperature or infrared radiation such as human sensors for toilet flushing systems or motion sensors. Night vision cameras are one of the applications of pyroimage sensors, which visualize a thermal (or infrared intensity) distribution image. Infrared light emitted from an object, which should be warmer than the surrounding temperature, is filtered with a germanium lens selecting an infrared beam (long wavelength), which is focused onto the pyroelectric target through an optical chopper. The temperature distribution of the object is represented on the target as a shrunk image, leading to a pyroelectric voltage distribution, which is monitored from the back surface of the target. Selection of the optimized chopping frequency is important, and for regular pyrosensors, it is around 1–100 Hz. Chopping too quickly decreases temperature perturbation, while chopping too slowly realizes the temperature saturation of the pyrosensor and thermal diffusion among the sensor pixels. Neither cases degrade the image quality. Figure 5.20 is an example of such a photo, taken during the Iraq War that started in 2003. You can clearly identify a horse and three soldiers in this frame.

5.5.2.3 *Surveillance Camera*

In collaboration with Penn State's International Center for Actuators and Transducers (ICAT), *QorTek* (Williamsport, PA), US, developed a solid-state gimbal system by coupling ultrasonic motors and the Gimbal operation mechanism to enable it to be small, lightweight, low power, and with almost no

Fig. 5.21.   Compact solid-state Gimbal mechanism with a noiseless ultrasonic motor zooming/focusing camera module.

moving parts. Most importantly, the motion generates no audible gear noise, like conventional electromagnetic motors. The system maintains a high image performance and provides precision-controlled high angular rate motion over a *wide acceptance angle*, which can be utilized for infrared and visible imaging on-missile seekers, search and tracking devices, and surveillance equipment. We adopted the then world-smallest camera module, and paired it with the ICAT micro-ultrasonic motor technology, which *Samsung Electromechanics*, Korea, utilized in the compact optical zoom camera modules for mobile phones [24]. As shown in Fig. 5.21, the basic construction of Solid State Gimbals is similar to a mouse trackball mechanism with an accurate 2D angle resolution, in which four piezoelectric strip-type linear ultrasonic motors are used for smooth and quiet ball motion (i.e., with vibrations at ultrasonic frequency — much higher than the human audible frequency range). Since the zero net charge (ZNC) drive was adopted, it has enabled a reduction in the number of components and total weight compared to conventional mechanical dual servo systems.

5.5.2.4 *Hand-Thrown Ball-Robots*

Hand-thrown type ball-robots were developed by NEC Corporation (Aerospace and Defense Operations Unit), Japan, in 2013, under the directive of Japan's Ministry of Defense. It appears to be a 5" diameter ball when folded up, but after it has been thrown by hand, it moves with its ball wheels which deploy when the ball arrives at the programed position, before finally developing its structure and camera for monitoring/surveying the situation — like a robotic "Transformer." Once a weapon, such as a compact laser gun, is installed in it, an ideal pin-point target weapon can be realized.

5.5.3 *Unmanned Vehicles*

Unmanned vehicles are vehicles without a person on board; that are either controlled remotely or are autonomous vehicles capable of sensing the environment and navigating on their own. These technologies are essential to reduce the number of military casualties. Though Japan was known as one of the best robot manufacturing countries, during the Tohoku Big Earthquake it was disclosed that the lack of such unmanned autonomous vehicles inevitably increased the death toll. Though Japan has a humanoid robot, ASIMO (acronym for Advanced Step in Innovative Mobility), developed by Honda, and a female android, Geminoid F, developed by Hiroshi Ishiguro, these technologies are mainly for entertainment — far from practical usage for crisis rescue.

5.5.3.1 *Unmanned Surface Vehicles*

Figure 5.22(a) shows an unmanned surface vehicle (USV), iRobot, which was supplied to the Japanese government for nuclear power plant surveillance immediately after the Fukushima accident in 2011. Though the Japanese people are still not interested in surveillance robots, after the company name "iRobot" became popular, sales of their vacuum cleaner USV product "Roomba" increased by more than 100% in less than one year.

5.5.3.2 *Unmanned Aerial Vehicles*

Unmanned Aerial Vehicles (UAV) have more than just military applications. The same machines whose missiles incinerate terrorists, can also help farmers apply water and pesticides to crops in a more precise manner — which can help save money and reduce environmental impacts. They can also help police departments find missing people, reconstruct traffic accidents, and act as lookouts for SWAT teams. They can alert authorities to people stranded on rooftops by hurricanes as well as monitor evacuation flows. States use them to inspect bridges, roads, and dams. Oil companies use them to monitor pipelines, while power companies use them to monitor transmission lines. See Fig. 5.22(b).

On the other hand, the civilian unmanned aircraft industry has begun to worry that, because of the public fear that the technology will be misused,

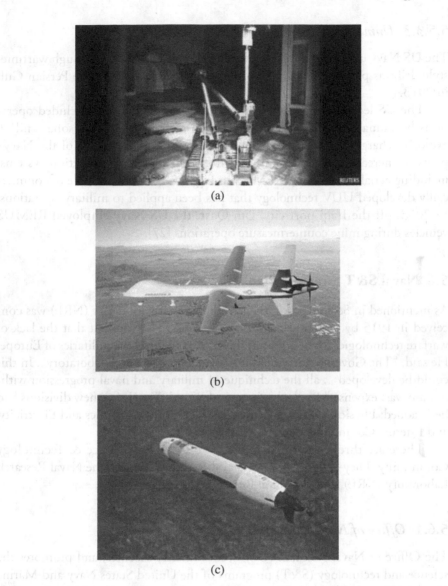

Fig. 5.22. (a) iRobot USV for the surveillance of the Fukushima Daiichi Power Plant. (b) Drone used for the battle field. (c) REMUS UUV during mine countermeasure operations.

strict regulations may happen before the adoption of the technology spreads widely. Another problem is the delay in the issuance of government safety regulations that need to come into force before drones can gain broad access to US skies [25].

5.5.3.3 *Unmanned Underwater Vehicles*

The US Navy used Unmanned Underwater Vehicles (UUVs) through wartime-style drills as part of international mine-clearing exercises in the Persian Gulf in 2013.

The US-led exercises, comprising more than 41 nations, included operations by unmanned Sea Fox devices, which are equipped with sonar and an explosive charge designed to shoot and destroy mines. It is part of the Navy's plans to increasingly deploy automated surveillance and protection systems, including aerial drones [26]. REMUS in Fig. 5.22(c) is an example of commercially developed UUV technology that has been applied to military operations. In 2003, off the Iraqi port city, Um Qasr, the US Navy employed REMUS vehicles during mine countermeasure operations [27].

5.6 Naval S&T Strategic Plan

As mentioned in Section 1.4.2, the Naval Research Laboratory (NRL) was conceived in 1915 by *Thomas Alva Edison* (1847–1931), who felt that the lack of warfare technologies in America put the USA far behind the militaries of Europe. He said, "The Government should maintain a great research laboratory... In this could be developed ... all the technique of military and naval progression without any vast expense." By the beginning of World War II, five new divisions had been added: Physical Optics, Chemistry, Metallurgy, Mechanics and Electricity, and Internal Communications.

There are three major divisions in the US Navy Science & Technology Community. They are the Office of Naval Research (ONR), the Naval Research Laboratory (NRL), and the Naval Research Enterprise (NRE).

5.6.1 *Office of Naval Research*

The Office of Naval Research (ONR) coordinates, executes, and promotes the science and technology (S&T) programs of the United States Navy and Marine Corps [28]. It holds a role like a "holding company" or "bank" in the whole Navy "Corporation." Its six subdivisions are coded from 30 to 35:

(Code 30) Expeditionary Maneuver Warfare and Combating Terrorism — Develops and transitions technologies to enable the Navy-Marine Corps team to win and survive on the battlefield.

(Code 31) Command, Control Communications, Computers; Intelligence, Surveillance and Reconnaissance — Supports the development of advanced electronics, communications, surveillance, and navigational tools that provide the Navy with a battlefield advantage.

(Code 32) Ocean Battlespace Sensing — Equips the Navy with technologies designed to observe, model, and predict air, ocean and shore environments, and detect underwater threats.

(Code 33) Sea Warfare and Weapons — Develops and delivers technologies that enable superior warfighting and energy capabilities for naval forces, platforms, and undersea weaponry.

(Code 34) Warfighter Performance — Enhances warfighter effectiveness and efficiency through bioengineered and bio-robotic systems, medical technologies, improved manpower, personnel, training, and system design.

(Code 35) Naval Air Warfare and Weapons — Supports the Navy's power projection needs, fostering the technology development of naval aircraft, structures, propulsion, autonomy, energetic, directed energy, and electric weapons.

As a SONAR component developer, I usually contacted Program Officers in Code 32 to submit an R&D proposal to ONR.

5.6.2 *Naval Research Laboratory*

The Naval Research Laboratory (NRL) holds a role similar to a "central research laboratory" in the whole Navy "Corporation." NRL operates as the Navy's full-spectrum corporate laboratory, conducting a broadly based multidisciplinary program of scientific research and advanced technological development directed toward maritime applications of new and improved materials, techniques, equipment, systems, and ocean, atmospheric and space sciences and related technologies [29].

After immigrating to the US, *Albert Einstein* (Professor at the Institute for Advanced Study, Princeton University) recommended to then-President *Franklin D. Roosevelt* to develop a nuclear bomb in the US (i.e., the *Manhattan Project*). Partially belonging to the NRL, Einstein helped with the development of the nuclear bomb, which was practically and effectively used in Hiroshima and Nagasaki, Japan, during World War II to end the war.

5.6.3 *Naval Research Enterprise*

The Naval Research Enterprise (NRE) holds a role similar to a "manufacturing division" in the larger Navy "Corporation" with many subdivisions.

- Naval Sea Systems Command (NAVSEA), the largest research division under the Navy's command, has the mission to design, build, deliver, and maintain ships and systems on time and on cost for the US Navy.
- Naval Undersea Warfare Center (NUWC) is a shore command of the US Navy within the NAVSEA Warfare Center Enterprise, which engineers, builds, and supports America's Fleet of ships and combat systems. As the Navy's premier research, development, test and evaluation (RDT&E) engineering, and Fleet support center for submarine warfare systems and other systems associated with the undersea battlespace, NUWC is charge of meeting the undersea warfare (USW) requirements of the 21st century.
- Naval Surface Warfare Center (NSWC) is another NAVSEA Warfare Center Enterprise. The NSWC provides the Navy's principal research, development, test and evaluation (RDT&E) assessment for surface ship systems and subsystems. In addition, the Warfare Centers provide depot maintenance and in-service engineering support to ensure the systems fielded today perform consistently and reliably in the future.
- Naval Air Systems Command (NAVAIR) has the mission to provide full life-cycle support of naval aviation aircraft, weapons and systems operated by Sailors and Marines. This support includes research, design, development, and systems engineering; acquisition; test and evaluation; training facilities and equipment; repair and modification; and in-service engineering and logistics support.
- Space and Naval Warfare Systems Command (SPAWAR), as the Navy's Information Dominance systems command, develops, delivers, and sustains communications and information capabilities for warfighters, keeping them connected anytime, anywhere. With space support activity, SPAWAR provides the hardware and software needed to execute Navy missions.

5.6.4 *Naval S&T Strategic Plan*

Here I introduce the "Naval S&T Strategic Plan" published by the Office of Naval Research in 2011. The contents described here have already been publicized in various international meetings by me in my capacity as a Navy officer until 2014, with the permission of ONR [30]. From its introduction, "The Naval Science and Technology (S&T) Strategic Plan describes how the Office of Naval Research (ONR) sponsors scientific research efforts that will enable the future operational concepts of the Navy and the Marine Corps. Department of Defense (DoD) and Department of the Navy (DON) strategic documents provide the basic foundation for this plan."

The Navy currently has nine focus areas (as elaborated in the subsections below), which highlight how S&T advances Naval capabilities and guides our investments.

5.6.4.1 *Assure Access to the Maritime Battlespace*

Assure access to the global ocean and littoral reaches and hold strategic and tactical targets at risk. Sense and predict environmental properties in the global ocean and littorals to support tactical and strategic planning and operations. Improve operational performance by adapting systems to the current and evolving environment (see Fig. 5.23).

5.6.4.2 *Autonomy and Unmanned Systems*

Achieve an integrated hybrid force of manned and unmanned systems with the ability to sense, comprehend, predict, communicate, plan, make decisions, and take appropriate actions to achieve its goals. The employment of these systems will reduce risk for Sailors and Marines and increase capability.

5.6.4.3 *Expeditionary and Irregular Warfare*

Naval warfighters of the future will possess the full spectrum of expeditionary kinetic and non-kinetic capabilities required to defeat traditional threats decisively and confront irregular challenges effectively.

Fig. 5.23. The Naval S&T Strategic Plan: assure access to the maritime battlespace.

5.6.4.4 *Information Dominance*

Enable the warfighter to take immediate, appropriate action at any time against any desired enemy, target or network by assuring that autonomous, continuous analyses of intelligence, persistent surveillance, and open information sources have, at all times, optimized the possible courses of action based on commander's intent.

5.6.4.5 *Platform Design and Survivability*

Develop agile, fuel efficient, and flexible platforms capable of operating in required environments. Enable manned and unmanned naval platforms and forces to operate in hostile environments while avoiding, defeating, and surviving attacks.

5.6.4.6 *Power and Energy*

Increase Naval forces' freedom of action through energy security and efficient power systems. Increase combat capability through high energy and pulsed power systems. Provide the desired power where and when needed at the manned and unmanned platform, system, and personal levels.

5.6.4.7 *Power Projection and Integrated Defense*

Enhance extended-range power projection capabilities and integrated layered defense by improving manned and unmanned Naval platforms, enabling forces to complete missions in hostile environments by avoiding, defeating, and surviving attacks. Demonstrate improvements in standoff indirect precision fires on time-critical targets, while limiting collateral effects through the use of electromagnetic kinetic projectiles, hypersonic missile propulsion, scalable weapons effects, directed energy, and hypervelocity weapons.

5.6.4.8 *Total Ownership Cost*

Support the goal to reduce Total Ownership Cost (TOC) by developing and aiding the insertion of technology to reduce platform acquisition cost, lifecycle and sustainment costs, and achieve crew manning requirements. TOC includes all costs associated with the research, development, procurement, operation, and disposal of platforms, combat systems, and associated elements over the full lifecycle.

5.6.4.9 *Warfighter Performance*

More effective point of injury care for Sailors and Marines. Enhanced health and warfighter performance both afloat and ashore. Highly efficient and effective human–system performance aided by new technologies created through the exploitation of biological design principles. Enhanced warfighter and system performance with reduced personnel costs as a result of the right information being provided to the right people with the right skills at the right time in the right jobs.

5.7 Summary

We considered crisis technologies in this chapter by adopting practical examples, including my contributions to piezoelectric actuators and sensors. I showed how electromagnetic field monitoring is promising for predicting earthquakes, and how anti-earthquake building structures with *shape memory alloys* (*SMAs*) were proposed. I showed how the MOV (*metal-oxide varistor*) with ZnO ceramic replaces a conventional lightening rod, and how *quartz membrane* can be used for various micro-mass sensors for monitoring epidemics/infectious diseases. I also showed how, in order to neutralize the biological weapon, a portable HClO disinfection device with a piezoelectric ultrasonic humidifier was developed. Regarding enormous accidents such as core meltdowns in nuclear power plant accidents, I showed how an Aluminum Nitride (AlN) piezoelectric transducer for monitoring the condition of the uranium rods in a nuclear chamber is definitely essential. In the intentional accident category, various antiterrorism technologies were introduced, including the *gamma-ray scan inspection system*, which can identify atomic species of liquids, i.e., differentiate liquid explosives from water, which is very essential at airport security check points. We talked about how the development of cyber weapons is one of the most recent military foci, to combat/prevent cyber threats and attacks. I also presented how, for warfare, *environmentally-friendly "green" weapons* became the mainstream focus in the 21st century in addition to the development of various unmanned aerial, surface, and underwater vehicles for surveillance and offensive purposes. Lastly, I presented the Naval S&T Strategies, as released by the Office of Naval Research.

5.8 Problem

A piezoelectric biosensor is fabricated using a piezoelectric PZT plate ($5 \times 1 \times 0.2$ mm^3), and driven at its fundamental resonance frequency f_R as shown in

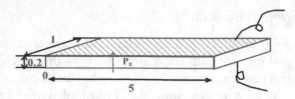

Fig. 5.24.   Piezoelectric resonator bio-sensor.

Fig. 5.24. When bacteria are adsorbed on the piezo-sensor uniformly, and the resonance frequency is slightly changed by Δf_R (the monitoring resolution of by $\Delta f_R/f_R = 10^{-6}$), evaluate the minimum weight amount of bacteria adsorbed on the sensor. Suppose that this frequency change is mainly attributed to the mass/density change, without changing the elastic constants or sensor size. The PZT's density is 8 g/cm³. [Hint] The fundamental resonance frequency is given by $f_R = v/2L = 1/2L\sqrt{\rho s_{11}^E}$ (v: sound velocity in PZT; L: length 5 mm; ρ: weight density; s_{11}^E: elastic compliance of PZT).

Taking into account the relationship $\ln(f) = \ln(l/2L) - (1/2)\ln(s^E) - (1/2)\ln(\rho)$, and primarily only the mass/density change, we obtain $\Delta M/M = \Delta\rho/\rho = 2\Delta f_R/f_R = 2 \times 10^{-6}$. Since $M = 8$ (g/cm³) $\times 5 \times 1 \times 0.2 \times 10^{-3} = 8 \times 10^{-3}$ (g), the mass detection resolution ΔM will be $2 \times 10^{-6} \times 8 \times 10^{-3}$ (g).

References

[1] http://www.glcoherence.org/monitoring-system/about-system.html.

[2] http://science.nasa.gov/science-news/science-at-nasa/2003/11aug_earthquakes/.

[3] J. Ryu, A. Vazquez Carazo, K. Uchino, and H. E. Kim, Magnetoelectric properties in piezoelectric and magnetostrictive laminate composites, *Japan. J. Appl. Phys.* **40**, 4948–4951 (2001).

[4] Y. Wang, D. Gray, D. Berry, J. Gao, M. Li, J. Li and D. Viehland, An extremely low equivalent magnetic noise magnetoelectric sensor, *Adv Mater.* **23**(35), 4111–4114 (2011).

[5] S. R. Debbarma and S. Saha, Review of Shape Memory Alloys applications in civil structures, and analysis for its potential as reinforcement in concrete flexural members, *Int'l J. Civil & Structural Eng.* **2**, 924–942 (2012).

[6] http://www.moriliving.com/en/estate/services/aseismic.

[7] http://www.fujitsu.com/global/about/resources/news/press-releases/2014/0414-01.html.

[8] http://en.wikipedia.org/wiki/Tornado.

[9] http://en.wikipedia.org/wiki/Varistor.

[10] http://www.medicinenet.com/swine_flu/article.htm#what_is_the_swine_flu.

[11] A. Pezeshk, Y. Gao, and K. Uchino, Ultrasonic piezoelectric hypochlorous acid humidifier for disinfection applications, *NSF EE REU Penn State Ann. Res. J.* **II**, ISBN 0-913260-04-5 (2004).

[12] http://www.reuters.com/article/2014/11/11/us-fujifilm-ebola-idUSKCN0IV0EJ 20141111.

[13] http://phillydeclaration.org/2014/04/15/thomas-jefferson-university-researchers-hope-to-fight-ebola-using-rabies-vaccine/.

[14] http://en.wikipedia.org/wiki/Three_Mile_Island_accident.

[15] B. R. Tittmann, Proc. ICAT 62nd International Smart Actuator Symposium, State College, PA, Oct 4–5 (2011).

[16] http://www.crimemuseum.org/crime-library/types-of-terrorism.

[17] http://www.yomiuri.co.jp/dy/national/T120102002799.htm 2/17/2012.

[18] http://www.reuters.com/article/2015/04/02/us-usa-cybersecurity-idUSKBN0MS 4DZ20150402.

[19] http://www.washingtonpost.com/news/checkpoint/wp/2014/12/10/with-photos-and-video-navy-shows-how-its-new-laser-gun-works-at-sea/.

[20] http://www.foxnews.com/tech/2015/02/05/us-navys-new-star-wars-style-railgun-hits-mach-6/.

[21] http://www.atk.com/MediaCenter/mediacenter_video gallery.asp.

[22] K. Uchino, Proc. New Actuator 2010, A3.0, Bremen, Germany, June 14–16 (2010).

[23] K. Uchino *et al.*, Deformable mirror using the PMN electrostrictor, *Appl. Opt.* **20**(17), 3077 (1981).

[24] K. Uchino, Piezoelectric motors for camera modules, *Proc. 11th Conf. Actuators*, Germany, p. 157 (2008).

[25] http://www.military.com/daily-news/2013/03/29/drone-industry-worries-about-privacy-backlash.html.

[26] http://www.military.com/daily-news/2013/05/15/us-navy-tests-anti-mine-drones-in-gulf-drills.html.

[27] http://www.navy.mil/navydata/cno/n87/usw/issue_26/uuv.html.

[28] http://www.onr.navy.mil/.

[29] http://www.nrl.navy.mil/about-nrl/mission/.

[30] *Naval S&T Strategic Plan*, published by Office of Naval Research (2011).

CHAPTER 6

Sustainability Technology

One type of politically initiated *normal technology* is *sustainability technology*. Sustainability technologies address the following concerns:

1. Power and energy (e.g., new energy harvesting devices and nuclear power plants, motivated by a perceived insufficient supply of oil)
2. Rare material (e.g., rare-earth metals, lithium)
3. Food (e.g., rice, corn — which also double as biofuel)
4. Toxic materials

   — Restriction (of, e.g., heavy metals, Pb, dioxin)
   — Elimination/neutralization (of, e.g., mercury, asbestos)
   — Discovery/creation of replacement material

5. Environmental pollution (management and mitigation)
6. Energy efficiency
7. Bio/medical engineering

In this chapter, I will discuss sustainability technology from the viewpoint of creating a sustainable society with a special emphasis on the area of piezoelectric devices.

6.1 Power and Energy

Based on the data reported by the *US Energy Information Administration* in 2009 [1], the energy sources for generating electricity in various countries such as (a) USA, (b) Japan, (c) China, (d) Russia, (e) France, (f) Germany, (g) Sweden, (h) Denmark, and (i) Brazil are shown in Fig. 6.1. The energy sources are classified as (1) hydroelectric, (2) fossil fuel (oil, coal, natural gas), (3) biomass, (4) nuclear, (5) windmill, (6) geothermal, and (7) solar cell. Although, at present, most major countries rely on fossil fuel and nuclear power plants, renewable energy sources have also been deployed, albeit in a minor way.

After the Fukushima Daiichi Nuclear Power Plant Accident in 2011, all nuclear power plants in Japan were expected to be shut down by 2015. If that had happened, 90% of Japan's electricity would have had to be generated using fossil fuels; which would have led to a dramatic jump in the amount of CO_2 generated by Japan — which goes against Japan's previous promise in the Kyoto Protocol. Japan will, in reality, need to restart the nuclear power plants as soon as they establish the necessary crisis technologies and management systems for the plants' facilities.

Some European countries, such as *Italy and Germany*, seem to be keen on gradually reducing their dependence on nuclear power plants; aiming instead, to increase their use of renewable energy sources. In contrast, *France, Sweden, Korea, USA, and Russia* are keen on retaining their nuclear power plants for the future. In Fig. 6.1, (g) Sweeden, (h) Denmark and (i) Brazil show a rather peculiar status. Sweden uses virtually no fossil fuels; it obtains half of its electricity from "dams," and the remaining from nuclear energy and biomass fuels. Denmark does not use nuclear energy at all and generates electricity from fossil and bio-mass fuels, with more than 17% generated by windmills (40% in 2014). To encourage investment in wind power, families were offered tax exemption for generating their own electricity within their own or an adjoining municipality. While this involved purchasing a turbine outright, more often families pur-chased shares in wind turbine cooperatives which in turn invested in commu-nity wind turbines. By 1996, there were around 2,100 such cooperatives in the country. Opinion polls show that this direct involvement has helped the popu-larity of wind turbines, with some 86% of Danes supporting wind energy compared to existing fuel sources [2]. Most of the electricity in Brazil is gener-ated by hydroelectric dams. Biomass (which, in Brazil, typically comprises alcohol from sugarcane) satisfies the remaining demand. Brazil is considered to be the world's first sustainable biofuels economy and the biofuel industry leader, with its sugarcane ethanol regarded as "the most successful alternative fuel to date." There are no longer any light vehicles in Brazil running on pure gasoline. Since 2007, the mandatory auto-fuel blend is 25% anhydrous ethanol and 75% gasoline (E25 blend) [3].

We review all energy sources from the sustainability viewpoint in the following sections.

6.1.1 *Fossil Fuel*

Fossil fuels are essentially the remains of plants or animals. They provide us with a source of non-renewable energy. *Oil, coal*, and *natural gas* are called 'fossil fuels' as they are formed by the decomposition of dead plant and animal matter buried deep in the earth's crust. These fossil fuels are then pumped from underground

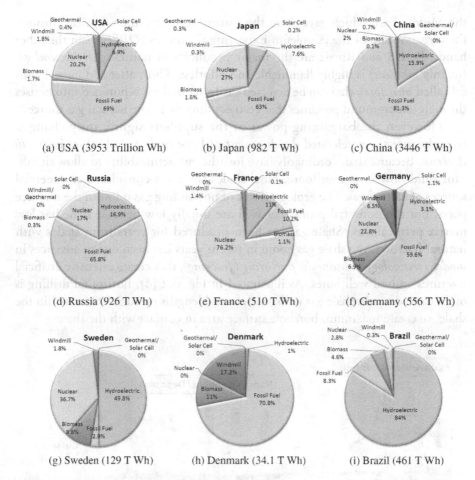

Fig. 6.1. Energy sources for generating electricity in various countries: (a) USA, (b) Japan, (c) China, (d) Russia, (e) France, (f) Germany, (g) Sweden, (h) Denmark, and (i) Brazil. Energy sources are classified into the following categories: Hydroelectric, Fossil fuel, Biomass, Nuclear, Windmill, Geothermal, and Solar cell. Data sources are based on the US Energy Information Administration Report 2009.

and used in a variety of ways. Since fossil fuels in their natural form must be burned in order to be used as electricity, they release inevitably unhealthy toxins into the air we breathe.

6.1.1.1 *Lack of Oil*

Because of the depletion of non-renewable sources, there could be a sudden and sharp decline in oil production in 30–100 years. The oil mines are geologically located in special areas. More than a third of the world's supply comes from the Middle East, which gave them the control to determine oil prices through the Organization of the Petroleum Exporting Countries (*OPEC*). On the other hand, the US and Russia are the major producers of natural gas. *Natural gas* (mainly *methane*) is highly flammable and odorless. Thus, after adding a chemical called *mercaptan* that can be detected easily, natural gas is pumped into houses through underground pipelines that connect directly to the natural gas source.

However, the bargaining power of the suppliers significantly changed recently after the accelerated development of large *shale gas*/oil mines in *North America*. Because shales ordinarily have insufficient permeability to allow significant fluid flow into a wellbore, most shales were not considered commercial sources of natural gas. The geological risk of not finding gas is low in the resource plays, but the potential profits per well are usually low because shale has low matrix permeability. Shale gas has been produced for years from shales with natural fractures; the shale gas boom in recent years has been due to advances in *modern technology in hydraulic fracturing (fracking)*, that create extensive artificial fractures around well bores. As illustrated in Fig. 6.2 [4], horizontal drilling is often carried out in shale gas wells, with lateral lengths up to 3,000 m within the shale, to create maximum borehole surface area in contact with the shale.

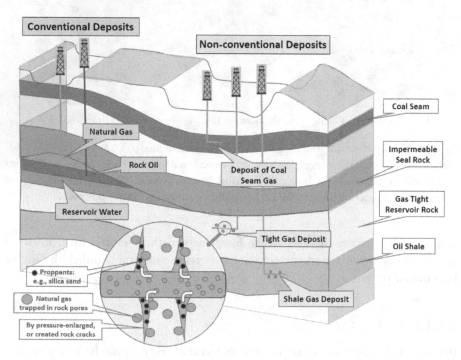

Fig. 6.2.   An illustration of shale gas compared to other types of gas deposits [4].

Shales are rich in organic material (0.5–25%). They are usually mature petroleum source rocks in the thermogenic gas window, where high heat and pressure convert petroleum into natural gas. Shales are also sufficiently brittle and rigid enough to maintain open fractures. The gas in the fractures is produced immediately; the gas adsorbed onto organic material is released as the formation pressure is drawn down by the well.

Another promising nonconventional source of natural gas is a *methane hydrate* mine discovered in the deep sea of the Pacific Ocean [5]. Methane hydrate is a cage-like lattice of ice inside of which are trapped molecules of methane. When methane hydrate is either warmed or depressurized, it reverts back to water and natural gas. When brought to the earth's surface, 1 cu. m of gas hydrate releases 164 cu. m of natural gas. In 2013, Japan became the first country to get gas flowing successfully from methane hydrate deposits under the Pacific Ocean. Hydrate deposits may be several hundred meters thick and generally occur in two types of settings: under Arctic permafrost, and beneath the ocean floor. Methane that form hydrates can be both biogenic (created by biological activity in sediments) and thermogenic (created by geological processes deeper within the earth).

After shutting down all nuclear power stations following the crisis at the Fukushima Daiichi (I) Nuclear Plant in 2011, Japan started expensive imports of liquefied natural gas from countries such as Qatar and the US. Thus, Japan has a big reason to pursue domestic methane hydrates. The territorial aggression over the Senkaku Islands (southern part of Japan) by China is presumably related to this attractive mine.

6.1.1.2 *Biofuel*

Unlike fossil fuels, *biofuels* are usually derived from recent biological materials, such as plants and animals. These fuels are made by converting biomass. Biomass can be converted into convenient energy-containing substances in three different ways: thermal, chemical, and biochemical conversions.

Bioethanol is an "alcohol" made by fermenting carbohydrates produced in *sugar or starch crops* such as corn, sugarcane, or sweet sorghum. Cellulosic biomass, derived from non-food sources, such as trees and grasses, is also being developed as feedstock for ethanol production. Ethanol can be used as a fuel for vehicles in its pure form, but it is usually used as a *gasoline additive* to increase octane and improve vehicle emissions. Bioethanol is widely used in the US and Brazil. Current plant designs do not cater for the conversion of the lignin portion of plant raw materials into fuel components by fermentation.

Biodiesel can also be used as a fuel for vehicles in its pure form, but it is usually used as a *diesel additive* to reduce levels of particulates, carbon monoxide, and hydrocarbons from diesel-powered vehicles. Biodiesel is produced from oils or fats through transesterification, and is the most common biofuel in Europe.

The production of biofuels also led to a flourishing automotive industry, where by 2010, 79% of all cars produced in Brazil were made to run on a hybrid fuel system of bioethanol and gasoline [3]. On the other hand, there are social, economic, environmental, and technical problems relating to biofuel production and use, which include: (1) the "food vs. fuel" debate, (see Section 4.1.3), (2) carbon emission levels, (3) sustainable biofuel production, deforestation and soil erosion, and (4) nitrous oxide (NO_2) emissions.

Although renewable sources of energy are better than non-renewable sources, we continue to use non-renewable sources because they are easier to obtain.

6.1.2 *Zero-Carbon Emission*

The Kyoto Protocol (see Section 3.3.3), which aims to regulate the amount of CO_2 being released into the atmosphere, is currently infringing on worldwide Electric Power Companies because a majority of the CO_2 generated per annum is attributed to electric power suppliers, as shown in Fig. 6.3 [6]. The Kyoto

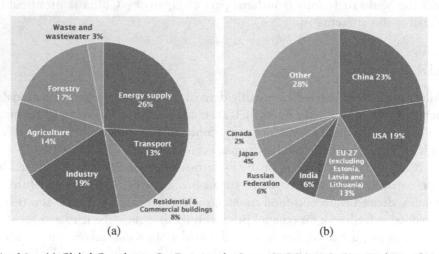

(a)   (b)

Fig. 6.3.   (a) Global Greenhouse Gas Emissions by Source IPCC (2007); EPA Disclaimer: based on global emissions from 2004. (b) Global CO_2 Emissions from Fossil Fuel Combustion and some Industrial Processes.

Source: 2008 National CO_2 Emissions from Fossil-Fuel Burning, Cement Manufacture, and Gas Flaring.

Protocol states that industrialized countries must reduce their greenhouse gas emissions by an average of 5% during 2008–2012 for the periods of unit commitment from the 1990 levels. However, owing to the Big Earthquake in 2011, Japan is behind in its emissions reduction, and still needs to reduce its emission levels by 6%. In order to meet this goal, Japan and other developed countries must replace existing thermal (firing) power generation methods with cleaner forms of energy generation. In the following sections, we discuss three types of power generation that do not emit carbon: (1) hydroelectricity, (2) nuclear power, and (3) renewable energy.

6.1.3 *Hydroelectricity*

Conventional hydroelectric turbines and dams are possible means through which countries could reduce CO_2 generation. An example of this is the Three Gorges Dam in the People's Republic of China discussed in Section 4.2.1. Built in 2012, it is the largest *hydroelectric power* station in the world. However, the dam has been a controversial topic both domestically and abroad. The Three Gorges Dam flooded archaeological and cultural sites and displaced some 1.3 million people, and continues to cause significant ecological changes, including an increased risk of landslides.

Another intriguing approach is to use "*ocean current*," which exists in sea depths 50–150 m deep and 200 m from the coast. American and European researchers are not focusing on this type of renewable energy because there is no strong ocean current in the US and Europe except at the tip of the peninsula of Florida. However, in countries like Japan, strong currents of speeds between 3–4 knots, with no change during all 365 days of the year do exist — e.g., the Kuroshio and Oyashio Currents around Japan. Because of the deep sea setting, the currents face no disturbance from typhoons.

Nova Energy Co., Ltd., Japan, is developing turbines to harvest energy from ocean currents. The turbines are connected to electricity generators by a system of rotating axle joints [7]. Figures 6.4(a)–6.4(c) show a hydro-turbine with a "tuna" shaped design, and how to install it on the holder to transfer the rotation to the electric generator and the future hydroelectric plant, respectively. Its tuna-shape enables the turbine to move on both vertical and horizontal axes, so that its body is always in line with the flow of the current. The rotating-axle system proved to be a success — the turbines change their positions to follow changes in the current, and are untroubled by passing debris. The experimental results verified 10 kW generation from an ocean current flow at a speed of 3–4 knots, with a turbine of 3 m diameter × 6 m length. Nova Energy plans to build a turbine with a 14-m long propeller, with a view to eventually

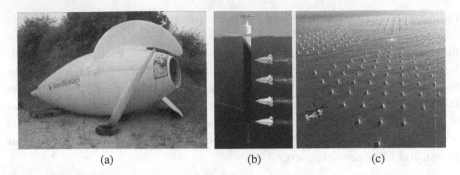

(a)   (b)   (c)

Fig. 6.4.   (a) Hydro-turbine with a "tuna" shaped design; (b) Installation of the turbine on the holder to transfer the rotation to the electricity generator; (c) concept drawing of an ocean current hydroelectric plant.

develop a 25 m-long model, which is expected to generate 500 kW per turbine. The 4 km × 4 km hydroelectric plant shown in Fig. 6.4(c) corresponds to one such nuclear power plant generating electricity.

Ocean currents have a continuous and unidirectional flow, unlike tidal waves. The setting of ocean current turbines like Nova Energy's in the deep sea is not expected to cause a negative impact on ocean traffic or on the marine ecosystem, which places it in a superior position to ocean windmill projects. Such an option is suitable for Japan.

6.1.4 *Nuclear Power Plant*

6.1.4.1 *Advantage/Disadvantage of Nuclear Power Plant*

When compared to oil-fired power generation, the main advantage of nuclear power is its reduced greenhouse gas emissions. The process of *nuclear fission* does not produce greenhouse gases, and therefore the only emissions by nuclear energy are produced during the transportation or extraction of energy from uranium. Since the shutdown of nuclear power plants in 2011, Japan has seen a vast increase in its CO_2 levels. An addition of about 100 million tons of CO_2 is released into the atmosphere per year than when the reactors were in operation — an added 8% to the country's total emissions. Aside from being a very clean source of energy, nuclear power facilities can produce energy up to a 91% efficiency rate.

However, as proved by the Fukushima catastrophe, the main disadvantage to nuclear power is the risk that it involves [8]. In a nuclear power plant, a small failure can result in a radioactive leak, which has the potential to harm thousands of human lives in countries with high population densities, like Japan. In order to combat this risk, Japan invested in Tsunami defenses (nicknamed "The Great Wall of Japan") to increase flooding protection. From 2011–2012, Japan also conducted stress tests on their nuclear plants. These tests simulated a range of accidents, to determine how the plants would stand up to such events. Another major issue with nuclear power in Japan is the production of nuclear waste. At the end of 2013, Japan's 10 power companies were reported to have 55,610 used fuel assemblies, weighing about 13,236 tons altogether, stored at 18 nuclear power plants. These companies occupied almost 55% of the available pool storage space. The final disadvantage of nuclear power is its cost. On the front end of the fuel cycle, Japan's nuclear fuel cycle is based on imported uranium that is subject to fluctuating price and supply. In addition, in order for Japan to restart its current reactors to generate power, the estimated cost stands at around US$700 million to $1 billion per unit, regardless of reactor size or age.

6.1.4.2 *Types of Nuclear Reactors*

Nuclear reactors are classified according to their type of nuclear reaction. All commercial power reactors are based on *nuclear fission*. They generally use *uranium* and its product *plutonium* as nuclear fuel, though a thorium fuel cycle is also possible. Depending on the energy of the neutrons that sustain the fission chain reaction, fission reactors can be divided roughly into two classes as follows [9].

Thermal reactors are the most common type of nuclear reactors. They use slow or thermal neutrons to keep up the fission of their fuel. They contain neutron moderator materials that are used to slowdown the fast moving neutrons until their neutron temperature is thermalized, i.e., until their kinetic energy approaches the average kinetic energy of the surrounding particles. Thermal neutrons have a far higher cross-section (probability) of fissioning the fissile nuclei, i.e., uranium-235, plutonium-239, and plutonium-241, and a relatively lower probability of neutron capture by uranium-238 (U-238) compared to the faster neutrons that originally result from fission, allowing the use of low-enriched uranium or even natural uranium fuel. The moderator is often also used as the coolant — usually water under high pressure to increase the boiling point. They are surrounded by a reactor vessel which is an instrument to monitor and control the reactor, a radiation shielding, and a containment building.

Fast neutron reactors use fast neutrons to cause fission in their fuel. They do not have a neutron moderator to slowdown the fast neutrons, and use less-moderating coolants. To keep the chain reaction under control, the fuel needs to be highly enriched in fissile material (about 20% or more) due to the relatively lower probability of fission vs. capture by U-238. Fast reactors have the potential to produce less transuranic waste because all actinides are fissionable with fast neutrons. However, they are more difficult to build and more expensive to operate. Therefore, due to these disadvantages, fast reactors are less common than thermal reactors in most applications. Some early power stations were used as fast reactors, as are some Russian naval propulsion units. The construction of prototypes is still ongoing.

On the other hand, *nuclear fusion* power is an experimental technology that generally uses hydrogen as fuel. Although not suitable for power production, Farnsworth–Hirsch fusors are used to produce neutron radiation.

6.1.4.3 *Fukushima Nuclear Power Plant Accident*

The Fukushima Daiichi (I) Nuclear Power Plant, operated by Tokyo Electric Power Company (TEPCO), is a disabled nuclear power plant located on a 3.5 km² site in the Futaba District of Fukushima Prefecture, Japan. This plant was commissioned in 1971, and consists of six boiling water reactors (BWRs). These light water reactors drive the electrical generators with a combined power of 4.7 GW, thereby making Fukushima Daiichi one of the 15 largest nuclear power stations in the world. The plant suffered major damage due to a serious earthquake that measured 9.0 on the Richter scale that led to a tsunami that struck Japan on March 11, 2011. This disaster disabled the reactor cooling systems, resulting in large radioactivity release into the environment and triggering a 30 km evacuation zone surrounding the plant. Note that the sister plant, known as the Fukushima Daini (II) Nuclear Power Plant (also run by TEPCO), located to the south did not suffer any serious accident during the tsunami, and cooling continued uninterrupted even after the disaster. We discuss in Chapter 7 that this difference happened mostly due to the crisis managerial process, rather than because of a technical problem in the TEPCO nuclear reactors.

The reactors of Fukushima Daiichi (I) Nuclear Power Plant's Units 1, 2, and 6 were supplied by *General Electric*, USA; Units 3 and 5 by *Toshiba*; and Unit 4 by *Hitachi*. All six reactors were designed by General Electric [10]. Units 1–5 were built with Mark I type containment structures, whereas Unit 6 has a Mark II type (over/under) containment structure. Unit 1 is a 460-MW BWR, constructed in 1967 (illustrated in Fig. 6.5). In February 2011, Japanese regulators granted an extension for a period of 10 years for the continued operation of the reactor — prior to it being damaged during the 2011 Tohoku earthquake and tsunami.

Unit 1 was designed for a peak ground acceleration of 0.18 G (1.74 m/s²), but rated for 0.498 G. The design basis for Units 3 and 6 were 0.45 G (4.41 m/s²) and 0.46 G (4.48 m/s²), respectively. All units were inspected after the 1978 Miyagi

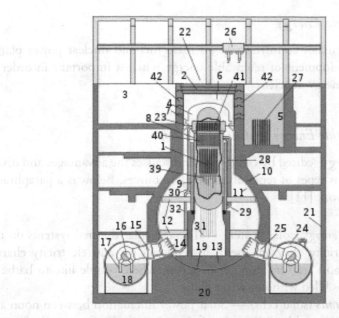

Fig. 6.5. Cross-section sketch of a typical BWR Mark I containment, as used in Fukushima Daiichi (I) Nuclear Power Plant. The reactor core (1) consists of fuel rods and moderator rods (39) which are moved in and out by the device (31). Around the pressure vessel (8), there is an outer containment (19) which is closed by a concrete plug (2). When fuel rods are moved in or out, the crane (26) will move this plug to the pool for facilities (3). Steam from the dry well (11) can move to the wet well (24) through jet nozzles (14) to condense there (18). In the spent fuel pool (5), the used fuel rods (27) are stored [Source; Wikipedia [10].].

earthquake when the ground acceleration was about 0.125 G (1.22 m/s^2) for 30 s, but the critical parts of the reactor were found to have hardly any faults. The design basis for tsunamis was 5.7 m.

The emergency diesel generators and DC batteries of the reactors are the crucial components in keeping the reactors cool in the event of power loss; these were located in the basements of the reactor turbine buildings. The design plans of the reactor provided by *General Electric* involved placing the generators and batteries in that location, but the Japanese mid-level engineers working on the construction of the plant were concerned that this made the backup power systems vulnerable to flooding. Unfortunately, TEPCO decided to strictly follow General Electric's design in the construction of the reactors. In the end, despite the main nuclear reactors enduring the large acceleration of the earthquake itself, it was the emergency energy supplies' location in the basements of the reactor turbine buildings that caused the critical accident.

6.1.5 *Renewable Energy*

At present, most major countries rely on fossil fuel and nuclear power plants; however, the development of renewable energy remains important in order to supplement the energy deficiency.

6.1.5.1 *Renewable Energy — Categorization*

"**Renewable Energy**" edited by *Godfrey Boyle* discusses the advantages and disadvantages of various types of renewable energy resources. Below is a paraphrased summary of his work [11]:

- *Solar thermal energy* (solar heater) — Though solar thermal systems do not generate electricity, they are useful in reducing gas and electricity charges dramatically, and are very popular in Japan, where people like to bathe in large bath tubs filled with hot water daily.
- *Solar photovoltaics* (solar cells) — Solar power fluctuation between noon and night is a major problem. Because of the relatively low energy density per area, a wide facility area is generally required. The state of Arizona in the US is suitable for solar cell farms as it does not have many rainy days in a year. However, this system is less recommended for Japan due to its many rainy or snowy days. Though the Japanese government subsidizes half the installation cost for solar panels, individual house owners still need to spend ~US$25,000 to set up a panel on their roof. From my personal experience, this investment will only break-even after 25 years, even though the electric charge in Japan is more than double of that in the US.
- *Biomass fuels* (thermal energy) — CO_2 generation is a major problem.
- *Tidal wave/Ocean current* hydroelectricity — Seasonal and daily fluctuations are the major problems. The major conflict is the disturbance of maritime activities; in particular, fishermen's boats in Japan. However, as mentioned in Section 6.1.3, deep *ocean currents* seem to be a promising type of renewable energy (no seasonal or daily fluctuation) for Japan.
- *Wind turbines* — Wind power fluctuation is the key to installation. The state of Pennsylvania is a windmill state in the US. An almost constant jet stream (latitude N40°) on the Appalachian Mountains (altitude around 1,000 m) allows constant power generation. This is less recommended for Japan because of typhoons that cause serious damage every year. The major disadvantage is the large size of wind turbines, which will likely interfere with maritime activities, as well as affect human residential areas on land, in Japan.

- *Geothermal* — Since Japan is a country with active volcanos, harvesting this form of energy also seems promising. Problems may exist in terms of personnel safety from volcanic activity.
- *Other forms of energy harvesting* — Piezoelectric and magnetoelectric effects can generate electricity from any noise vibration, such as those produced by automobiles or infrastructures (e.g., in/by buildings, on/by bridges, etc.).

6.1.5.2 *Selection Criteria*

The selection criteria for the prospective renewable energy suitable for each country/state or local area are given below:

- Reasonable power level
- Harvesting efficiency
- Steady harvesting (consider daily/seasonal changes and weather-dependency)
- Maintenance (in consideration of environmental effects, such as snow, storms, tornados, etc.)
- State political issues (e.g., The Japanese government hesitates to work on new methods which have not yet been tried in other countries. This is the major reason why Japan is currently behind in producing renewable energy technologies compared to the US and Europe, as seen in Fig. 6.1).

6.1.5.3 *Solar Cell*

Solar cell roofs in residential areas have become rather popular in Japan nowadays. Since one system set typically costs US$25,000; taking into account the average annual electricity saving amount (which we assume to be ~US$1,000), the break-even point is obtained at around 25 years. Because most Japanese live in their houses for more than 30 years, even after his/her retirement (remember that the average expected lifetime for the Japanese female is now 87 years), this initial investment may be returned to the last dollar — assuming that the individual is not a victim of any major accidents and/or natural disasters. In the US, since people frequently move (I change my residence every seven years or so), this investment is not very beneficial to them.

Regarding Mega/Giga Watt (solar) projects in Japan, the government invested largely on the infrastructure in northern Hokkaido Island. However, due to snow and lack of power transmission cables, most of the projects have either crashed or have been delayed. The presence of sunshine throughout the year is also essential for the project's success — as seen in Arizona, USA.

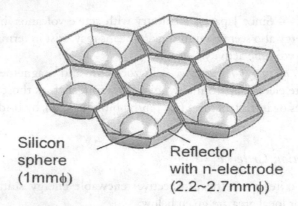

Fig. 6.6.   Solar cell design and alignment in 2D.

For portable power supply applications, the low efficiency and crystal fragility of silicon cells are the major problems. Hoping to solve these problems, *CleanVenture 21*, Osaka, Japan, has been working on Silicon Solar Cells that are both flexible and unbreakable [12]. Figure 6.6 shows the design of the solar cell with silicon spheres (each sphere's diameter measures 1 mm ϕ or less) aligned on an aluminum bed with parabolic concaves. This design pro-vides the following three benefits:

- Reduction in the amount of silicon material used; leading to a reduction in total weight and price.
- Increase in effectiveness of light irradiation due to the reflection of light by the parabolic concaves.
- Increase in flexibility and breaking tolerance of the solar cell panel.

The trend of declining customer interest in installing solar modules on the roofs of their factories, warehouses and facilities has been said to be due to the current models on the market being too heavy. This design (Fig. 6.6) addresses the problem with its light-weight and flexible modules — each weighing less than 5 kg due to the spherical silicon cells.

6.1.5.4 *Tidal Wave/Ocean Current*

Tidal power electricity generation projects pursued by other countries, such as Britain and South Korea, have been troubled by frequent occurrences of debris, such as driftwood and fishing nets, jamming or breaking turbine propellers.

Since the problem is primarily due to the fixed position of the turbines, Nova Energy's "Tuna" turbines (Fig. 6.4) solve the issue with their ability to move freely, and follow changes in the direction of the *deep ocean currents* they harvest energy from. This allows debris to wash past the turbines rather than get lodged in their propellers.

6.1.5.5 *Wind Turbine*

There are three primary types of wind turbines, as shown in Fig. 6.7 [13].

Horizontal-Axis Wind Turbines (HAWTs): In HAWTs, the main rotor shaft and electrical generator are located at the top of a tower, and must be pointed into the wind. Small turbines are pointed by a simple wind vane, while large turbines generally use a wind sensor coupled with a servomotor. Most of these turbines have a gearbox, which turns the slow rotation of the blades into a quicker rotation that is more suitable to drive an electrical generator [Fig. 6.7(a)]. This type of turbine is most popularly found in the United States and Japan.

Vertical-Axis Wind Turbines (VAWTs): VAWT Savonius types have the main rotor shaft arranged vertically [Fig. 6.7(b)]. One advantage of this arrangement is that the turbine does not need to be pointed into the wind to be effective, which is an advantage on a site where the wind direction is highly variable. It is also an advantage when the turbine is integrated into a building because its placement causes it to be inherently less steerable. Also, the generator and gearbox are located near the ground, using a direct drive from the rotor assembly to the ground-based gearbox — thus improving accessibility for maintenance. Its key disadvantages are as follows: relatively low rotational speed with the consequential higher torque and hence the higher cost of the drive train; the inherently lower power coefficient; the 360 degree rotation of the aerofoil within the

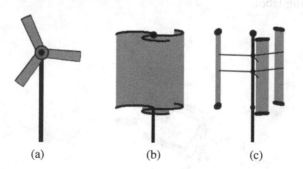

(a)   (b)   (c)

Fig. 6.7.   Three primary types of wind turbines: (a) HAWT towered, (b) VAWT Savonius, and (c) VAWT Darrieus.

wind flow during each cycle, which causes high dynamic loading on the blade; the pulsating torque generated by some rotor designs on the drive train; and the difficulty of modeling the wind flow accurately — hence the challenges of analyzing and designing the rotor prior to fabricating a prototype.

Darrieus Turbines: VAWT Darrieus types, named after their inventor *Georges Darrieus*, have good efficiency, but produce a large torque ripple and cyclical stress on the tower, which contributes to poor reliability [Fig. 6.7(c)]. They generally require some external power source or an additional Savonius rotor to start turning, since the starting torque is very low. The torque ripple is reduced by using three or more blades, which results in the greater solidity of the rotor. Solidity is measured by dividing blade area by the rotor area. Newer Darrieus type turbines are not held up by guy-wires but have an external superstructure connected to the top bearing.

6.1.5.6 *Piezoelectric Energy Harvesting*

The harvesting of piezoelectric energy is one recent research interest. In the energy harvesting process, a cyclic electric field excited in the piezoelectric plate by environmental noise vibrations is accumulated into a rechargeable battery. This can be seen in an LED traffic light array system developed by NEC-Tokin, Japan. The system is driven by a piezoelectric windmill that operates using the wind generated by passing automobiles. Other successful piezoelectricity-driven products (determined as products sold by a minimum of one million sellers) in the commercial market include the "Lightning Switch" (a remote switch for room lights using a unimorph piezoelectric component) developed by PulseSwitch Systems, VA, USA [14]. In addition to living convenience, the Lightning Switch (Fig. 6.8) also reduces the construction costs of houses drastically, due to a significant reduction in the amount of copper electric wires used and the aligning labor.

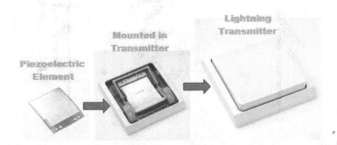

Fig. 6.8.   Lightning switch with piezoelectric thunder actuator. [Courtesy of Face Electronics.]

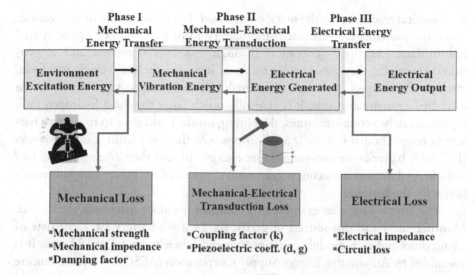

Fig. 6.9. Three major phases associated with piezoelectric energy harvesting): (i) mechanical–mechanical energy transfer, (ii) mechanical–electrical energy transduction, and (iii) electrical–electrical energy transfer.

There are three major phases/steps associated with piezoelectric energy harvesting [15]: (i) mechanical–mechanical energy transfer, which involves the mechanical stability of the piezoelectric transducer under large stresses, and mechanical impedance matching, (ii) mechanical–electrical energy transduction, which relates with the electromechanical coupling factor in the composite transducer structure, and (iii) electrical–electrical energy transfer, which includes electrical impedance matching, such as a DC/DC converter that helps in converting the energy into a rechargeable battery (see Fig. 6.9). The Penn State group developed energy harvesting piezoelectric devices based on a "Cymbal" structure (29 mmφ, 1–2 mm thick), which can generate up to 100 mW of electric energy under an automobile engine's vibration [16]. By combining three cymbals in a rubber composite with a washer-like energy harvesting sheet (developed for a hybrid car application), the system aims to enable a constant accumulation of 1W-level electricity in a Li-ion battery (connected to the energy harvesting system).

6.1.6 *Fuel Cell/Batteries*

Electricity generated by the above-mentioned renewable technologies should be stored in either fuel cells or batteries for future usage. Fuel cells and batteries are similar since both use a chemical reaction to generate electricity. A battery stores the chemical reactants; usually metal compounds like lithium, zinc or manganese, Once the reactants are exhausted, the battery must be recharged again. A fuel cell initiates electricity generation through reactants (*hydrogen* and *oxygen*) stored externally; they generate electricity as long as a fuel cell vehicle is *refueled* instead of recharged. Although different kinds of hydrogen tanks and ways to store hydrogen are available, it is generally much faster to refill a hydrogen tank (approximately a couple minutes, depending on the tank) than to recharge a battery (a couple hours) in vehicle applications. On the other hand, *Electric Vehicles* (EV) with batteries are two-times more energy-efficient than *Hydrogen Fuel Cell Vehicles*, and the instantaneous acceleration or speed of the power supply is much better for the batteries.

The Nissan Leaf is an example of a compact five-door hatchback electric car. Manufactured and introduced in 2010, its 24 kWh battery pack consists of 48 modules and each module contains four cells, thereby a total of 192 cells. It is assembled by Automotive Energy Supply Corporation (AESC) — a joint venture between Nissan, NEC, and NEC Energy Devices. According to Dr. Kazuaki Utsumi (Executive Expert, AESC, Zama, Japan), NEC's battery technology is a "Laminate-type Mn–Li-ion battery," i.e., a large current from the lamination is possible, and Mn–Li-ion batteries are much safer than Co–Li-ion types. Owing to the volumetric shrinkage in $LiCoO_2$ or $LiNiO_2$ batteries, these batteries sometimes cause thermal explosions, while $LiMn_2O_4$ has not caused any explosive incidents so far (Private Communication).

In contrast, major automobile companies like Toyota Motors, Japan, Hyundai Motors, Korea, and Honda Motors, Japan, are seeking hydrogen fuel cells. Interestingly, Tesla Motors, USA, like Nissan, uses lithium-ion batteries. The winner in this contest may determine which technology (batteries or fuel cells) powers our vehicles for decades to come.

6.1.7 *Electricity Grid*

Depending on the nature of the energy source an electric power station obtains its power from (e.g., thermal, hydroelectric, windmill, solar, etc.), the stations' generative power levels deviate over significant periods (whether in a matter of hours, days, months, or even years) in addition to the differences in their supply voltage (i.e., either AC or DC). Thus, finding a way to combine different sources of electric power effectively is an essential problem. As mentioned earlier, the generation capability of tidal wave turbines is at its maximum only for a duration of six hours in the daytime, while solar cells do not operate during snowy winter seasons (on top of their usual daily modulation).

A traditional *electricity grid* is an interconnected network for delivering electricity from suppliers to consumers. It consists of power stations that generate electrical power, high-voltage transmission lines that carry power from distant sources to demand centers, and distribution lines that connect individual customers [17]. Power stations may be located near a fuel source, e.g., at a dam site, to take advantage of renewable energy sources, and are often located away from heavily populated areas. They are usually quite large, so as to take advantage of the economies of scale. The generated electric power is stepped up to a higher voltage at which it connects to the electric power transmission network. The bulk power transmission network will transmit power over long distances, sometimes across international boundaries, until it reaches its wholesale customer (usually the company that owns the local electric power distribution network). On arrival at a substation, the power will be stepped down from a transmission level voltage to a distribution level voltage. After it exits the substation, it enters the distribution wiring. Finally, upon arrival at the service location, the power is stepped down again from the distribution voltage to the required service voltage(s).

In contrast to traditional electrical grids, which are used to carry power from a few central generators to a large number of users or customers, a new emerging *smart grid* is now being developed. It uses two-way flows of electricity and information to create an automated and distributed advanced energy delivery network. A smart grid allows the power industry to observe and control parts of the electric system at higher resolution in time and space. It allows for customers to obtain cheaper, greener, less intrusive, more reliable, and high quality power from the grid. The legacy grid did not allow for real time information to be relayed from the grid, so one of the main purposes of the smart grid is to allow *real time information* to be received and sent from and to various parts of the grid, to make operation as efficient and seamless as possible.

Electricity must be generated on demand and the voltage and frequency of the AC supply have to be held within relatively tight limits. Various technologies are being used to meet unexpected increases in demand. When the demand is low, surplus electricity can be used to pump water into high level reservoirs. At times of sudden peak demand, they use the stored potential energy of the water to generate electricity. Rechargeable batteries have also been used by electricity utilities for peaking power and emergency backup since the late 19th century. The largest power system (40 MWh) is located in California. It allows us to manage the logistics of the grid and view consequences that arise from its operation on a time scale with high resolution — from high-frequency switching devices on a microsecond scale; to managing wind and solar output variations on a minute scale; to managing the future effects of the carbon emissions generated by power production on a decade scale.

6.2 Rare Material

6.2.1 *Rare-Earth Metals*

In Section 4.1.2, we discussed rare materials in terms of rare-earth metals and lithium. China decreased the production of rare-earth material by 40% in 2010. This was done to drive up their price, because China currently dominates the rare-earth industry. This strategy of global economic manipulation is similar to OPEC's international "cartel" on oil prices. In order to produce electric vehicles, hybrid cars and wind turbines, a strong magnet with *non-rare-earth metals* at a (preferably low) cost unaffected by China's rare-earth production control is needed. To meet this need, *Don Heiman* at Northeastern University, Massachusetts, developed a magnetic material with a manganese gallium compound, which is very strong. This compound is synthesized on a nanoscale to produce a strong magnetic field that rivals that of rare-earth magnets, which are considerably more expensive to process and maintain.

6.2.2 *Rare Metals*

The internal electrode material of choice for conventional ceramic multilayer (ML) devices were Ag/Pd, as it was in capacitors. However, owing to the Russian economic crisis in the late 1990s, the price of palladium increased dramatically, peaking in 2000 at a price which was 10 times higher than its price in 1990 (Fig. 6.10). Because of this change in raw material cost, capacitor industries started to use Ni (base metal) internal electrodes instead. Though the price of

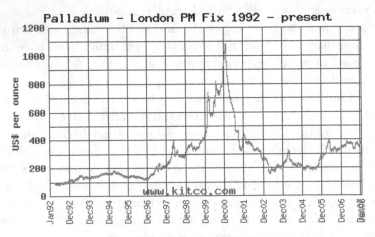

Fig. 6.10.   Palladium price change with year.

Pd has stabilized now, cheaper materials are still being used. This is an example how the industrial "Threat" transformed into an "Opportunity" for products. The necessity created the technology.

The piezo-actuator industry is a little behind in using Cu internal electrodes widely. Piezo-industries are still using Ag/Pd or Pt for internal electrodes. Only recently have manufacturers like TDK-EPCOS, Austria, started to commercialize Cu embedded ML actuators for diesel injection valves, in which the actuator cost is the present major bottleneck [18].

6.3 Food/Water

6.3.1 *Food*

As explained in Section 4.2.3.1, famine is defined as a widespread scarcity of food, caused by several factors including crop failure, population imbalance, or government policies. This phenomenon is usually accompanied or followed by regional malnutrition, starvation, epidemics, and increased mortality.

Long-term measures to improve food security include investing in modern agricultural techniques (such as fertilizers and irrigation), and strategic national food storage measures. In addition, recent work has indicated that the entire human population could be fed in the event of a global catastrophe if alternative foods were used.

Artificial fertilization systems using a piezoelectric actuator can impregnate cows precisely and quickly, at the rate of some 200/h. Because of the high responsivity of the piezo-actuator in comparison to a normal hydraulic oil pressure system, sophisticated artificial fertilization systems were developed by Applied Micro Systems, Japan [19]. Figure 6.11 compares the difference in egg deformation during the needle insertion process for the new piezo-actuator system against that of the conventional oil-pressure system. It is obvious that the deformation can be minimized and the needle insertion can be made much quicker by using piezo-actuation, since it has the ability to superpose a 100 Hz high frequency motion onto the needle. This equipment may solve the issue of starvation in Africa.

6.3.2 *Water*

Water is more crucial for survival than all other resources on earth. However, over 1 billion men, women, and children on the planet today still drink unsafe water. The lack of access to safe drinking water is one of the primary causes of hunger, disease, and poverty throughout the developing world.

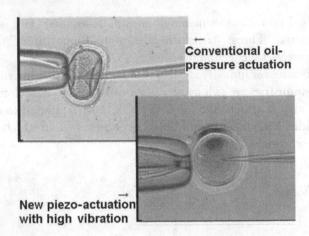

Fig. 6.11.   Egg deformation during the needle insertion process using the piezo-robot usage (bottom) *vs.* the conventional oil-pressure system (top). [Courtesy of Applied Micro Systems.]

Without water, crops and livestock wither and die. People go hungry and become weak. Weakness allows diseases to run their course until they finally become *Quiet Killers*. At this moment, many communities throughout the world are suffering needlessly because water is either scarce, polluted or may not exist at all. Underground water availability, though so near, remains too far for people lacking the tools and knowledge to reach it. The lack of safe drinking water is also the primary cause of disease (80% of health problems) in the world today. Every day, tens of thousands of people die due to diseases that are directly related to contaminated water. And for those who survive without good health, there is very little chance of them leading a normal and productive life. Much of this is because, in rural areas of developing countries, the only source of water for domestic purposes is often either a badly polluted and shallow well (less than 10 feet deep), or a mud-hole used by both animals and humans.

By 2025, it has been estimated that half of the world's population will lack access to safe drinking water. Although roughly 70% of our planet is covered with water, potable water is becoming increasingly scarce as the world's population and the demand for water continue to grow. According to what science knows today, there is enough freshwater on the planet for 7 billion people; however, the freshwater is distributed unevenly and too much of it is wasted, polluted, and unsustainably managed. Many of these water-based challenges can be addressed with appropriate water treatment solutions. BASF, USA, has developed "Sokalan®," an antiscalant which acts as a scale control dispersant (the equipment that desalts the water from sea water [20]) (see Fig. 6.12). Modern desalination facilities do so using the distillation process, but due to the

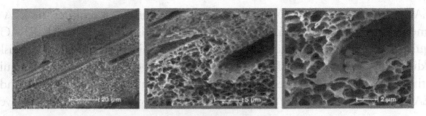

Fig. 6.12. Scanning electron microscope pictures of antiscalant membrane. [Courtesy of BASF.]

high consumption of energy when heating the water, an alternative method using specific membranes developed in the 1970s to remove impurities and make the water pure, was invented. The *reverse osmosis process* is used to purify water through membrane filtration and is commonly employed in sectors such as paper, food, steel, industrial chemicals, pharmaceuticals, automotive, and petrochemicals industries. Nowadays, both methods are used, and desalinated seawater has become one of the major alternatives to ground water; helping to overcome the problem of water scarcity in dry regions.

6.4 Toxic Material

6.4.1 *Restriction of Toxic Materials*

6.4.1.1 *TSCA*

The Toxic Substances Control Act (TSCA) passed in the US in 1976 was developed to fill in the gaps of previously existing laws, all of which were designed to protect human health and the environment from chemicals released by industry and by humans each day (such as through driving, flushing the toilet, and even taking pharmaceuticals). Under TSCA, the Environmental Protection Agency (EPA) evaluates the toxic effects of new and existing chemicals on both humans and the environment. It also mandates the tracking of the thousands of chemicals that are produced or imported into the country. Although TSCA mentions the testing and assessment of individual chemicals and mixtures, there are no apparent guidelines for the assessment of mixtures. Additionally, when referring to mixtures, the language of TSCA seems to refer to commercial or industrial mixtures [21].

6.4.1.2 *REACH*

In 2006, the European Union (EU) enacted REACH (Registration, Evaluation and Authorization of Chemicals), a new approach to chemical regulation.

REACH is essentially a mirror image of TSCA. While TSCA requires the EPA to demonstrate that a chemical is a risk to human or environmental health, REACH requires that manufacturers test and ensure that chemicals do not pose such risks. Additionally, under REACH, many old or existing chemicals will also require testing, unlike TSCA. Since chemicals imported into the EU also fall under REACH, the legislation may impact some chemicals produced in the US as well.

6.4.1.3 *Restrictions on the use of certain Hazardous Substances*

RoHS, short for "Directive on the restriction of the use of certain hazardous substances in electrical and electronic equipment," was adopted in February 2003 by the EU, as introduced in Section 3.3.3. RoHS applies to the products in the EU that are listed in the next paragraph, regardless of whether they are made within the EU or imported. Though this is a local restriction, most of the manufacturers of electronic components are sensitive toward it as long as they are exporting their products to EU countries.

RoHS restricts the use of the following six substances:

- Lead (Pb)
- Mercury (Hg)
- Cadmium (Cd)
- Hexavalent chromium (Cr^{6+})
- Polybrominated biphenyls (PBB)
- Polybrominated diphenyl ether (PBDE).

6.4.2 *Replacement Materials*

Can the piezoelectric industry survive in the 21st century without using lead zirconate titanate (PZT)? With RoHS limiting the usage of lead (Pb) in electronic equipment, we may need to regulate the usage of PZT, which is the most popular piezoelectric ceramic at present. Japanese and European communities may experience governmental regulation on PZT usage in the next 10 years. Lead-free piezoceramics were first developed after 1999, and are classified into three types: $(Bi,Na)TiO_3$, $(Na,K)NbO_3$, and tungsten bronze (also known as BNT, NKN and TB, respectively).

6.4.2.1 *BNT*

The share of the patents for bismuth compounds (bismuth layered type and $(Bi,Na)TiO_3$ (*BNT*) type) exceeds 61%. This is because bismuth compounds are easily fabricated in comparison to other compounds. Note that the toxicity of Bi^{3+} is not very low compared to Pb^{2+}; however, a regulation may not be proposed because the atomic element of Bi^{3+} is not (yet) familiar to politicians. The Langevin transducers popularly used for underwater fish finders and hydrophones are also utilized for ultrasonic cleaning and hazardous material dissolving systems. For these ecological applications, it is reasonable to also make the systems environmentally-friendly. *Honda Electronics*, Japan, developed Langevin transducers by using BNT-based ceramics for ultrasonic cleaner applications [22]. The composition $0.82(Bi_{1/2}Na_{1/2})TiO_3–0.15BaTiO_3–0.03(Bi_{1/2}Na_{1/2})$ $(Mn_{1/3}Nb_{2/3})$ O_3 exhibits $d_{33} = 110 \times 10^{-12}$ C/N contains only 1/3 of a hard PZT, but its electromechanical coupling factor $kt = 0.41$ is larger because of a much smaller permittivity ($\omega = 500$) than that of PZT. Furthermore, the maximum vibration velocity of a rectangular plate (k_{31} mode) is close to 1 m/s (rms value), which is higher than that of a typical hard PZT.

6.4.2.2 *NKN*

$(Na,K)NbO_3$ (*NKN*) systems exhibit the highest performance because of the morphotropic phase boundary usage. Figure 6.13 shows the current best data reported by *Toyota Central Research Lab*, where strain curves for oriented and unoriented (K, Na, Li) (Nb, Ta, Sb)O_3 ceramics are shown [23]. Note that the maximum strain reaches up to 1500×10^{-6}, which is equivalent to the PZT strain. The major drawbacks of NKN systems include their sintering difficulty and the necessity for the sophisticated preparation technique (*the topochemical method*) for preparing flaky raw powder.

Tungsten–bronze (*TB*) types are another alternative choice for resonance applications, due to their high Curie temperature and low loss. Taking into account the general consumer attitude on the disposability of portable equipment (e.g., printers, cellular phones), *Taiyo Yuden*, Japan, developed micro-ultrasonic motors using non-Pb multilayer piezo-actuators [24]. Their composition is based on TB $((Sr,Ca)_2NaNb_5O_{15})$ without heavy metals or even K (potassium seems to be a little toxic in comparison to Na). Though the basic piezoelectric parameters in TB ($d_{33} = 55$–80 pC/N, $T_C = 300°C$) are not very attractive, once the c-axis oriented ceramics are prepared, the d_{33} is dramatically enhanced up to 240 pC/N. Further, since the Young's modulus $Y_{33}^E = 140$ GPa is more than twice of that of PZT, the higher generative stress is expected, which is suitable for ultrasonic motor applications. *Taiyo Yuden* developed a sophisticated preparation technology for fabricating *oriented ceramics* with a multilayer configuration: preparation under a strong magnetic field — much simpler than the flaky powder preparation in NKN. Since most piezo-ceramics are diamagnetic, the

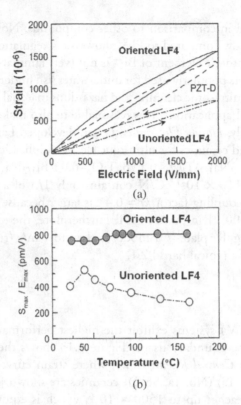

Fig. 6.13. Strain curves for oriented and unoriented (K, Na, Li) (Nb, Ta, Sb) O₃ ceramics.

ceramic powder suspended in slurry will be aligned along its magnetically-stable axis under a strong magnetic field (such as 10 T). Since the polarization axis in their particular TB composition corresponds to the magnetically-unstable axis, they used the magnetic field parallel to the green sheet. A cut-green-sheet was rotated practically under a steady magnetic field during the drying period.

Figure 6.14(a) shows their compact rotary ultrasonic motor with a piezoelectric multilayer actuator (MLA) [24]. A cantilever rod is wobbled by a 2×2 arrayed MLA element, driven by a four-phase voltage (sine, cosine, –sine, and –cosine). They fabricated monolithic 2×2 arrayed elements with the size $2.6 \times 2.6 \times 1.1$ mm³ — including buffers with layer thickness as thin as 18 μm [see Figs. 6.14(b) and 6.14(c)]. Because of this thin layer and a dramatic enhancement in the piezoelectric performance owing to the *magnetic field alignment*, the ultrasonic motor was successfully driven under only 3 V_{p-p}, which is low enough to be adopted in mobile phones without coupling a step-up drive circuit.

Refer to a review paper on Pb-free piezoelectrics by Tsurumi [25] to obtain further information. Though their piezoelectric performance is not superior to

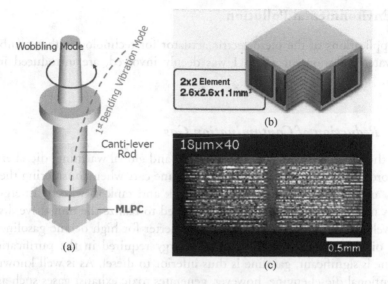

Fig. 6.14. (a) Compact rotary ultrasonic motor with a piezoelectric multilayer actuator, (b) 2 × 2 arrayed element, and (c) cross-section view of the 18 μm-layer ML element [22].

the PZTs, commercialization efforts have already been made. These are introduced in the following section on ultrasonic cavitation cleaners by *Honda Electronics*, Japan.

6.4.3 *Elimination of Hazardous Materials*

Ultrasonic lens cleaners are commonly used in homes, while industrial ultrasonic cleaners are widely utilized in the manufacturing lines of Silicon wafers and liquid crystal display glass substrates. Cavitation (vacuum particle) is generated in water by increasing the ultrasonic power level in water. Because the cavitation (cyclic adiabatic compression at around 28 kHz) generates higher than 3000°C micro-locally for a short period, we can make hazardous waste innocuous. Various hazardous wastes produced in the late 20th century, including dioxin, trichloroethylene, polychlorinated biphenyl (PCB), environmental hormones, etc., can be found underground or in sewer water. It is well known that dioxin attains the properties of a toxic material when it is burned at the low temperatures typical in garbage disposal furnaces (~700°C), while it becomes innocuous only when burned at a very high temperature (~3000°C). Therefore, ultrasonic irradiation is highly prospected for this application. Since the chemical reaction is induced in water (i.e., so-called *sonochemistry*), dioxin can be dissolved into an innocuous material without an apparent increase in the water temperature.

6.5 Environmental Pollution

The applications of the piezoelectric actuator for technologies that combat air and water pollution, in which I was deeply involved, are introduced in this section.

6.5.1 *Reduction of Contamination Gas*

From the viewpoint of energy conservation and global warming, diesel engines are more energy-efficient than regular gasoline cars when considering the total energy of gasoline production (well-to-tank and tank-to-wheel). Energy efficiency, measured by the total energy required to realize the unit drive distance for a vehicle (MJ/km), is of course much better for high octane gasoline than diesel oil. However, since the electric energy required in the purification of gasoline is significant, gasoline is thus inferior to diesel. As is well known, the conventional diesel engine, however, generates toxic exhaust gases such as SO_x and NO_x, which decreases its efficiency. In order to solve this problem, new diesel injection valves were developed by Siemens, Bosch, and Toyota with piezoelectric multilayer actuators. Figure 6.15 shows a common rail-type diesel injection valve [26].

In order to eliminate toxic SO_x and NO_x and increase the efficiency of diesel engines, high pressure fuel and quick injection control are required. The highest reliability of these devices at an elevated temperature (150°C) for a long period

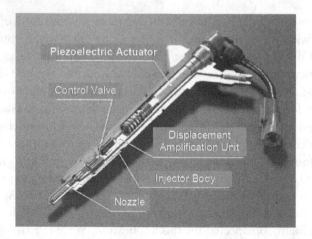

Fig. 6.15.   Common rail type diesel injection valve with a piezoelectric multilayer actuator. [Courtesy of Denso Corporation.]

(10 years) has been achieved [26]. Note that the piezoelectric actuator is a key component for increasing the burning efficiency and minimizing the toxic exhaust gas elements.

6.5.2 *Protection of River/Sea Pollution*

Honda Electronics, Japan, added an ultrasonic stain remover into a washing machine produced by Sharp. Figure 6.16 shows the L–L coupler horn developed by Honda Electronics to generate water cavitation for removing oily dirt from a shirt collar. It is noteworthy that the amount of detergent needed per wash (which is one of the major causes of river contamination) can be significantly reduced by this technique.

6.6 Energy Efficiency

6.6.1 *Energy Density*

Electric components, such as motors and transformers, are mostly based on electromagnetic transduction at present. By reducing the size of these components (power level less than 30 W), their efficiency reduces drastically due to the *Joule heat* in their thin coil wire (i.e., the resistivity in the coil becomes significant). Note that the efficiency of a wristwatch motor for driving hour/minute/second hands is less than 1% (95% of the input battery energy is consumed by generating heat),

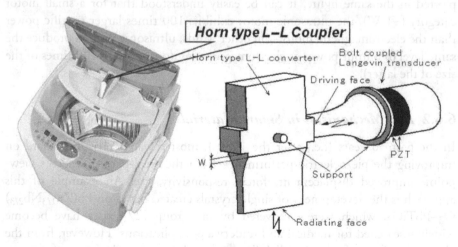

Fig. 6.16.   L–L coupler horn for a washing machine application. [Courtesy of Honda Electronics.]

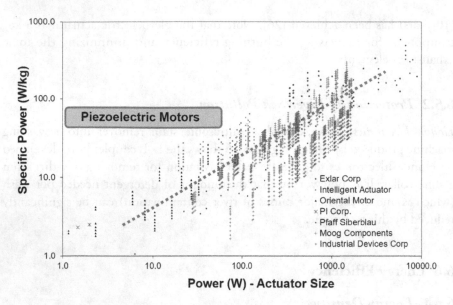

Fig. 6.17.   Power (W: actuator size) vs. specific power (W/kg) for various electromagnetic motors.

so that the hands move intermittently, as intended. Thus, piezoelectric actuators and transducers with much lower losses are highly sought after in the 21st century. Figure 6.17 plots the power level (W: actuator size) vs. specific power (W/kg: efficiency) for various electromagnetic motors. Note that the *specific power* reduces drastically with each reduction in motor size. The specific power for piezoelectric ultrasonic motors (~500 W/kg), which is insensitive to the motor size, is also plotted in the same figure. It can be easily understood that for a small motor category (~1 W), the ultrasonic motor exhibits 100 times larger specific power than the electromagnetic motors (in other words, ultrasonic motors produce the same level of output power as electromagnetic motors at just 1/100 times of the size of the latter).

6.6.2 *Loss Mechanisms in Smart Materials*

In the past 30 years (i.e., after the 1980s), most researchers worked hard on improving the piezoelectric performance from the "real-part" property's viewpoint: improved displacement, force, responsivity, etc. An example of this approach is the development of single crystals created based on $Pb(Zn_{1/3}Nb_{2/3})O_3$–$PbTiO_3$, which were discovered by my group [27]; they have become widely researched for medical and underwater applications. However, from the viewpoint of efficiency and reliability, such as heat generation or performance degradation under high voltage/power drive, the key is the *imaginary-part*, i.e., the *loss and hysteresis mechanisms*. Thus, since the beginning of the 2000s, I have been studying loss mechanisms in piezoelectrics, including high-power characterization systems (HiPoCS™) [28], and practical high-power piezo-ceramics, in addition to highly energy-efficient compact piezoelectric actuators and transducers.

6.7 Bio/Medical Engineering

These days, piezoelectric actuators have become rather widely used in medical applications. A summary of the advantages of piezoelectric devices over electromagnetic (EM) motors is listed below:

(a) More suitable for miniaturization — Since the stored energy density of the piezo-device is larger than that of an EM type, the desired volume and weight for micromotors (the spec requirement from the medical field currently aspires toward motors around 1–2 mm in size) can be achieved with as low as 1/10 of their original form, thus making them more compact.

(b) No electromagnetic noise generation — Since magnetic shielding is not necessary, the compact design can be maintained. This unique characteristic is particularly desirable for motors that are used under a high magnetic field, such as MRI equipment; where large piezo-motors can be used for rotating a patient or equipment, and small piezo-motors for microsurgery under MRI monitoring.

(c) Higher efficiency — The significant decrease in the efficiency of the EM motor is mainly due to the increase in Joule heat resulting from the reduction in coil wire thickness. Since the efficiency of the piezo-device is generally unaffected by its size, it is effective in the power range lower than 30 W. This characteristic is another attractive point for battery-operated portable and wearable medical devices. The same battery, when used in a piezoelectric motor as opposed to an EM motor, is expected to last more than 10 times longer.

(d) Non-flammable — Piezo-devices are safer in the event of an overload or short-circuit at the output terminal. This is again a benefit for portable and wearable devices.

Piezoelectric devices for medical applications are introduced below. The areas covered are an artificial fertilization system, a medical micropump, micro-monitoring and surgery, and drug delivery devices [29].

6.7.1 *Artificial Fertilization System*

Applied Micro Systems, Japan, commercialized a micro-robot (shown in Fig. 6.18) consisting of four electromagnets (for clamping) and four multi-layer actuators (for translation). Currently, the principle driver is "inchworm" motion on a steel surface with a 30 nm movement resolution. Micro-tools such as micro tweezers, scissors, and drillers are available in conjunction with the robot base, for manipulating nanoparticles and fibers.

The high responsivity of the piezo-actuator in comparison to a normal hydraulic oil pressure system, led to the development of sophisticated artificial fertilization systems. Figure 6.11 shows the egg deformation difference during the needle insertion process using a micro-robot vs. a conventional oil-pressure system. It is obvious that the deformation can be minimized by using piezo-actuation, since a high frequency motion of 100 Hz can be superposed onto the needle.

6.7.2 *Medical MicroPump*

PZT thin films are deposited on a silicon wafer which is then micro-machined to leave a membrane for fabricating micro-actuators and sensors, i.e., microelectro

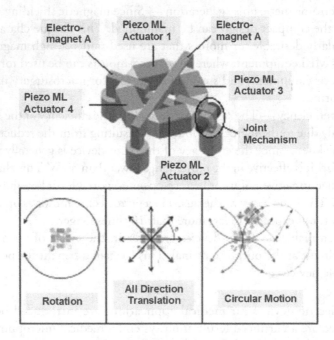

Fig. 6.18.   Micro-robot with four electromagnets and four ML actuators. [Courtesy of Applied Micro Systems.]

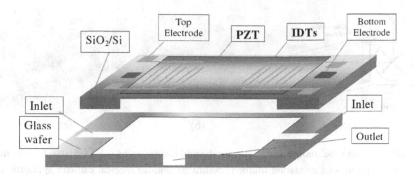

Fig. 6.19.   Structure of a PZT/silicon MEMS blood testing device.

mechanical systems. Figure 6.19 illustrates a blood tester prototype developed by Penn State University in collaboration with OMRON Corporation, Japan, in the 1990s. On applying voltage to two surface interdigital electrodes, the surface PZT film generates surface membrane waves, which soak up blood and the test chemical from the two inlets, then mix them in the center part, before transferring the mixture to the monitor part through the outlet. An FEA calculation was conducted to evaluate the flow rate of the liquid by changing the thickness of the PZT or the Si membrane, the size of the inlet and outlet nozzles, and the cavity's thickness; and the performance was verified experimentally.

Note that we also need thick PZT films for medical devices, because a certain volume of PZT is required to generate the required mechanical force/power. PZT can generate mechanical power in the range of 30 W per 1 cm^3 volume (max); so thin PZT films of 1 μm thickness are useless from the actuator viewpoint, though they are widely used as piezoelectric sensors, such as for automobile force/pressure/acceleration sensors.

6.7.3 *Micro-Monitoring and Surgery*

My research group at Penn State University is developing various micro-rotary motors (1–5 mm in size). A micromotor called the "metal tube type" consisting of a hollow metal cylinder and two (or four) PZT rectangular plates is used as a basic micro-actuator [see Fig. 6.20(a)]. When we drive one of the PZT plates, Plate X, a bending vibration is excited basically along the x-axis. However, because of an asymmetrical mass (Plate Y), another hybridized bending mode is excited with some phase lag along the y-axis, leading to an elliptical locus in a clockwise direction, in a hula-hoop-like motion. The rotor of this motor is a cylindrical rod with a pair of stainless ferrule pressed with a spring. The assembly is shown in Fig. 6.20(b). The metal cylinder motor, 2.4 mm in diameter and

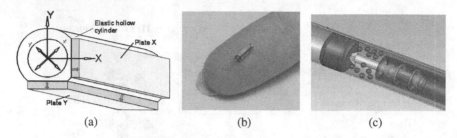

(a)   (b)   (c)

Fig. 6.20.   "Metal tube" motor using a metal tube and two rectangular PZT plates. (a) Schematic structure, (b) photo of a miniature motor (1.5 mm φ), and (c) medical catheter application for blood clot removal.

12 mm in length, was driven at 62.1 kHz in both rotation directions. A no-load speed of 1800 rpm and an output torque up to 1.8 mNm were obtained for rotation in both directions under an applied rms voltage of 80 V. A rather high maximum efficiency of about 28% for this small motor is a noteworthy feature.

In collaboration with Samsung Electromechanics, Korea, Penn State developed the world's smallest camera module with both optical zooming and autofocusing mechanisms for a cellular phone application in 2003. Two micro-ultrasonic motors with a 2.4 mm diameter and 14 mm length were installed to control zooming and focusing lenses independently in conjunction with screw mechanisms.

Micromotors less than 1.3 mm φ with 3 mm in length are very useful for microsurgery or minimally invasive surgeries. For example, Micromechatronics Inc., USA, in collaboration with Penn State University, developed the "Micro-Disrupter™" by combining the metal tube motor with a drill, as illustrated in Fig. 6.20(c). Coupled with a medical catheter, the drill could disrupt 'pig' blood clots successfully, which may help *brain cerebral infarctions* or even *stroke* or *heart-attack* patients on an ambulance.

6.7.4 *Drug Delivery*

High power ultrasonics are applicable to transdermal drug delivery. Penn State researchers are working on the commercialization of such a "needle-free" injection system for insulin by using compact but powerful "Cymbal" piezo-actuators. Figure 6.21(a) shows a transdermal insulin drug delivery system with four small piezoelectric cymbal transducers, and Fig. 6.21(b) pictures its test procedure with a mouse. The design of the device is based on research results that suggest that exposure to ultrasound of 100 mW/cm^2 for 5–10 min seems to be sufficient for insulin to penetrate the skin, leading to a dramatic decrease in glucose levels.

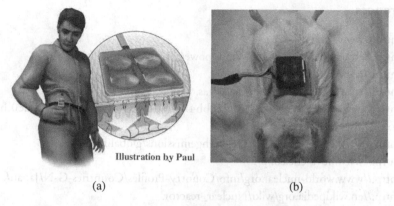

Illustration by Paul

(a)   (b)

Fig. 6.21.   Transdermal insulin drug delivery system (a) and its test procedure with a mouse (b).

6.8 Summary

Though renewable energy currently plays a very small role in global energy supplies in general, the following technologies continue to be developed: *Solar thermal energy, Solar photovoltaic, Biomass fuels, Tidal wave/Ocean current hydroelectricity, Wind turbines* and *Geothermal energy*. Other energy harvesting techniques being pursued include piezoelectric and magnetoelectric energy; for generating electricity from noise vibrations from automobiles or infrastructures.

This chapter showed how the dearth of rare materials, such as rare-earth metals and lithium, became a serious bottleneck for Japanese component industries; and how artificial fertilization systems with piezoelectric actuators can impregnate cows precisely and quickly, at the rate of nearly 200/h. I also explained that toxic materials are strictly regulated in the 21st century — that mercury and asbestos should be eliminated and neutralized, and heavy metals, such as Pb or Dioxin are restricted. As a result, Pb-free piezoelectric ceramic development is a hot research topic worldwide. We also discussed how, to solve the problems of environmental pollution, multilayer piezoelectric actuators were adopted in diesel engine injection valves to eliminate toxic SO_x and NO_x and increase the efficiency of diesel engines. This chapter also mentioned how piezoelectric motors exhibit higher energy efficiency than conventional electromagnetic motors for compact sizes below 30 W, in general. Thus, various micro- and energy-efficient bio/medical engineering devices using piezoelectric rather than electromagnetic motors are being developed. Examples of such devices include artificial fertilization systems, medical micropumps, micro- monitoring and surgery devices, and drug delivery instruments.

References

[1] http://www.eia.gov/.

[2] http://en.wikipedia.org/wiki/Wind_power_in_Denmark.

[3] http://en.wikipedia.org/wiki/Biofuel.

[4] http://en.wikipedia.org/wiki/Shale_gas.

[5] http://www.ft.com/cms/s/2/8925cbb4-7157-11e3-8f92-00144feabdc0.html#axzz3X8sgxBs4.

[6] http://www.epa.gov/climatechange/ghgemissions/global.html.

[7] http://www.nova-ene.co.jp/.

[8] http://www.world-nuclear.org/info/Country-Profiles/Countries-G-N/Japan/.

[9] http://en.wikipedia.org/wiki/Nuclear_reactor.

[10] http://en.wikipedia.org/wiki/Fukushima_Daiichi_Nuclear_Power_Plant.

[11] *Renewable Energy*, 3rd edn., Godfrey Boyle, (ed.), Oxford University Press (2012).

[12] http://www.cv21.co.jp/en/.

[13] http://en.wikipedia.org/wiki/Wind_turbine.

[14] K. Uchino and T. Ishii, Mechanical damper using piezoelectric ceramics, *J. Jpn. Ceram. Soc.* **96**(8), 863–867 (1988).

[15] K. Uchino, Energy harvesting with piezoelectric and pyroelectric materials, in *Energy Flow Analysis in Piezoelectric Harvesting Systems*, Chp. 4 — Patial Change, Nantakan Muensit (ed.), Trans Tech Publications, Zuerich, Switzerland (2011).

[16] H. W. Kim, S. Priya, K. Uchino, and R. E. Newnham, Piezoelectric Energy Harvesting under High Pre-Stressed Cyclic Vibrations. *J. Electroceram*, **15**, 27–34 (2005).

[17] https://en.wikipedia.org/wiki/Electrical_grid.

[18] F. Boecking and B. Sugg, *Proc. New Actuator* 2006 (Bremen, June 14–16), A5.0, 171 (2006).

[19] http://www.applied-micro-systems.net/en/a000.html.

[20] http://www.performancechemicals.basf.com/ev/internet/watersolutions/en/desalination/reverse-osmosis.

[21] http://www.eoearth.org/view/article/155695/.

[22] T. Tou, Y. Hamaguchi, Y. Maida, H. Yamamori, K. Takahashi, and Y. Terashima, Properties of $(Bi_{0.5}Na_{0.5})TiO_3$-$(Bi_{0.5}Na_{0.5})(Mn_{1/3}Nb_{2/3})O_3$ Lead-free piezo- electric ceramics and its application to ultrasonic cleaner. *Jpn. J. Appl. Phys.* **48**, 07GM03 (2009).

[23] Y. Saito, H. Takao, T. Tani, T. Nonoyama, K. Takatori, T. Homma, T. Nagaya and M. Nakamura, Lead-free piezoceramics, *Nature* **432**, 84–7 (2004).

[24] Y. Doshida, *Proc. 81st Smart Actuators/Sensors Study Committee*, JTTAS, Dec. 11, Tokyo (2009).

[25] T. Tsurumi, Development of Pb-free piezoelectrics, *J. Japan. Ceramic Soc.* **40**, (2005).

[26] A. Fujii, *Proc. Smart Actuators/Sensors Study Committee*, JTTAS, Dec. 2, Tokyo (2005).

[27] J. Kuwata, K. Uchino and S. Nomura Dielectric and piezoelectric properties of 0.91Pb(Zn1/3Nb2/3)O3-0.09PbTiO$_3$ single crystals, *Jpn. J. Appl. Phys.* **21**, 1298 (1982).

[28] K. Uchino, Y. Zhuang, and S. O. Ural, Loss determination methodology for a piezoelectricceramic: new phenomenological theory and experimental proposals, *J.Adv. Dielectrics* **1**(1), 17–31 (2011).

[29] K. Uchino, *Nice Little Movers*, Medical Device Developments, www.mdd-spg.com (March, (2006), p. 16).

CHAPTER 7

Risk Management

Regarding the Fukushima Nuclear Power Plant accident, there were many problems discussed at several stages. The first involved the prediction of earthquakes and tsunamis. Indeed, the capabilities of science to provide an accurate prediction of when earthquakes and tsunamis may strike is presently limited. In addition, there aren't sufficient monitoring and public reporting systems available. The second problem revolves around the initial actions that were taken after the nuclear accident. The Fukushima nuclear power plant meltdown occurred as a result of the electric shutdown that followed the unexpected large earthquake/tsunami. The crucial difference between the Three Mile Island accident in Pennsylvania in 1979 and the Fukushima accident in 2011 was the explosion at the Fukushima facilities that had occurred due to a delay in the cooling process, which was seen as the major failure in the initial action after which the resulting meltdown led to this difference. In this chapter, we explore why the delay in decisions and directives of the political leaders happened. The third is the order of the commanding system; the worst situation was caused by the political leaders' lack of knowledge on science and technology, in particular, on risk management. It was difficult for Tokyo Electric Power Company (TEPCO), a private company, to handle the crisis by itself, because it neither had nor used the robots necessary, and it did not have the ability to respond to the crisis. I found it difficult to digest the fact that the delay in the Chief Commander's decision brought about the explosion — and this is the reason I insist on the urgent necessity of the development of *crisis technology* in Japan. These were my major motivations to write this book.

In this chapter, we discuss the risk management issues for the (1) individual, (2) a non-nation/state or local area, and (3) a nation/state. The highest level of risk management issues, the international/global level, is discussed in the next chapter (Chapter 8).

7.1 Level of Risk Management

"Security" means the protection of the elementary value of a human, as an individual or a group, from various threats. Table 7.1 summarizes various situations

Table 7.1.   Various situations for different levels of threat provider and threat receiver.

Threat Receiver	Threat Provider				
		Non-Nation Group		Non-Human	
	Nation/State	Domestic	International (Refugee, Terrorist)	(Disaster, Disease)	Security Category
Nation/State	Traditional security	Rebellion, civil war	Terror attack on the state	Earthquake HIV, Bird-flu	Nation security treaty
Non-Nation Group	Racial oppression	Racial conflict	Social instability by refugees	Tsunami refugee	Racial/ company security
Individual	Human right suppression	Abduction by coup group	Abduction by terrorists	Death by disease, earthquake	Human security
Example	"Axis of Evil"	Peace maintenance	International crime	Environmental strategy	

according to the levels of threat provider and threat receiver [1]. The threat providers are classified into three categories: (1) nation/state, (2) non-nation groups (e.g., domestic rebels, international terrorists, etc.), and (3) non-human (e.g., natural disasters, infectious diseases, etc.), and the threat receivers are categorized into three categories: (1) nation/state, (2) non-nation groups, and (3) individuals.

7.1.1 *Threat Provider View*

Let us consider Table 7.1 in detail by viewing the table vertically.

7.1.1.1 *Nation/State*

When a nation declares a threat on another nation, this is a matter of *traditional security*. Such threats include both military methods and non-military methods such as economic or cultural means. Nations/states can also threaten domestic non-nation groups and/or domestic individuals' *human rights*.

After 1959, when the Dalai Lama's government had fled to India, China governed western and central Tibet as the *Tibet Autonomous Region*, while the eastern areas are now mostly ethnic autonomous prefectures within the Sichuan province. There are tensions surrounding Tibet's political status and between the

Chinese government and dissident groups which are active in exile. It is also said that Tibetan activists in Tibet have been arrested or tortured. In 1980, General Secretary and reformist Hu Yaobang visited Tibet, and ushered in a period of social, political, and economic liberalization. At the end of the decade, however, analogous to the *Tiananmen Square* protests of 1989, monks from the Drepung and Sera monasteries started protesting for independence, and so the government halted reforms and started an anti-separatist campaign [2]. Human rights organizations have been critical of Beijing's and the Lhasa government's approaches to human rights in the region where crackdowns on separatist convulsions around monasteries and cities have occurred — most recently in the 2008 Tibetan unrest.

The state of human rights in the Islamic Republic of Iran has also been criticized by Iranian and international human rights activists, writers, and NGOs. The United Nations General Assembly and the Human Rights Commission have condemned prior and ongoing abuses in Iran in published critiques and several resolutions. The government of Iran has been criticized both for restrictions and punishments that follow the constitution and the law of the Islamic Republic, and for actions that do not, such as the torture, rape, and killing of political prisoners, and the beatings and killings of dissidents and other civilians.

7.1.1.2 *Non-Nation Group*

Occasionally, a domestic non-nation group rebels against their nation — sometimes even overturning the government in the process. The *Arab Spring* is an example of such an event. It was a revolutionary wave of demonstrations and protests (both non-violent and violent), riots, and civil wars in the Arab world that began in December 2010 in Tunisia with the Tunisian Revolution, before spreading throughout the countries of the Arab League and its surroundings. The Arab Spring has often been described as a wave of popular uprisings against an oppressive rule (*Intifadas*). While the wave of initial revolutions and protests faded by mid-2012, some started to refer to the succeeding and present large-scale discourse conflicts in the Middle East and North Africa as the *Arab Winter*. The most radical discourse from the Arab Spring that continued on into the present civil wars, took place in Syria during the second half of 2011 [3].

Violent conflicts also occur between domestic non-nation groups, exemplified by the battle between the *Shia and Sunni Islamic sects*. Over the years, Sunni–Shia relations have been marked by both cooperation and conflict. Sectarian violence is a major element of friction throughout the Middle East and continues to cause problems in Pakistan and Yeman. Tensions between communities intensified during the various power struggles, such as the Bahraini uprising, the Iraq War, and most recently, in the Syrian Civil War and in the formation of the self-styled *Islamic State of Iraq and Syria* (ISIS) and its advancement on Syria and Northern Iraq.

International non-nation groups which provide a threat include international terrorist groups, pirates, and human traffickers. Al-Qaeda is a radical Islamic group organized by Osama bin Laden in the 1990s to engage in terrorist activities. They attacked the US on September 11, 2001; crashing hijacked commercial air- planes into both towers of the World Trade Center (private sector building) and the Pentagon (government institute). The September 11 attack resulted in the death of more than 2,600 people. Above national security, terrorists and human traffickers also threaten individual security. Since these violent non-nation groups do not follow global norms/regulations, we need to establish a suitable global security regime and system in order to defend against and/or counteract these attacks.

A sudden influx of a large number of *refugees* also creates instability in societies. A migrant influx crisis is considered a *social security* problem. Migrants from countries such as Syria, Afghanistan, and Africa have been entering through Greece and passing through Macedonia, Serbia, Hungary, and Croatia in hopes of reaching safety in Germany, which has opened its doors to receive refugees. Since these people had faced intolerable conditions in their homelands, they keep searching for the possibility of a better life with steady, decent work, readily available basic necessities like food and water, and educational opportunities for their children. But above all, they long for a life without the constant threat of war. However, countries around the world are currently facing a huge dilemma regarding the refugee situation; between accepting refugees, owing to humanitarian obligations/commitments, or rejecting refugees because of their potential threat to national security. The issue arose from the discovery of "fake refugees" (i.e., in reality, terrorists trained by ISIS) amongst genuine asylum seekers, after the attack on central Paris, France, in November 2015, where an estimated 130 people were killed, and 368 people injured in the coordinated attacks that took place. Amongst the dead were 89 victims of the terror attack at the Bataclan theater; where terrorists took hostages before engaging in a standoff with police.

7.1.1.3 *Non-Human (Disaster, Disease)*

Non-human factors such as natural disasters and infectious diseases also threaten the safety of the collective human group. If global warming continues to raise sea levels, some countries, such as Japan and the Netherlands, will find their territorial land areas drastically decreased. And as exemplified by the Kanto-Tohoku Earthquake on March 11, 2011, earthquakes, volcanic activities, and tsunamis damage national territories and kill many people at the same time. Then there is also the current outbreak of Ebola in West Africa (first cases reported in March 2014) — the largest and most complex outbreak since the Ebola virus was first discovered in 1976. There have been more victims in this outbreak than all the others combined owing to its spread out of West Africa — starting in Guinea, then spreading across land borders to Sierra Leone and Liberia, then by air to Nigeria and USA, and by land to Senegal and Mali.

7.1.2 *Threat Receiver View*

Now, we consider Table 7.1 by viewing it horizontally.

7.1.2.1 *Individual Level*

An individual may receive a threat from his/her nation/state. For example, activists in countries such as China and North Korea, where freedom of speech may be restricted, face the possibility of being arrested and/or exiled for their cause(s). An individual may also face threats from a foreign state — such as in the case of the *abduction of Japanese citizens* from Japan by agents of the North Korean government during a period of six years from 1977 to 1983. Although only 17 Japanese (eight men and nine women) have been officially recognized by the Japanese government as having been abducted, there may have been hundreds of victims. The North Korean government has officially only admitted to abducting 13 Japanese citizens. There have also been testimonies regarding the many non-Japanese citizens, including nine European citizens, who may have been abducted by North Korea.

People can also be threatened by both "external" (international) and "internal" (i.e., domestic) terrorists; as exemplified by the 9/11 attack on the World Trade Center in 2001 (see Section 7.1.1.2), and the Boston Marathon Bombing on April 15, 2013, in the USA. In the latter case, two domestic terrorists set a pair of homemade bombs a short distance from the finish line of the Boston Marathon. The bombs detonated in the crowd watching the race, killing three people and injuring more than 260.

In order to protect the individual, the "Anti-terrorist Training Program" in Section 7.3 was created.

7.1.2.2 *Non-Nation Group Level*

Non-nation groups, such as a particular minority race in a nation, may face the suppression of certain activities by the government or be threatened by another competitive race. Or, the government may threaten or destroy the natural environment in the home area of the minority race.

An example of this is the conflict between Tibet and China on March 17, 2008, as reported by *The Wall Street Journal* [2]: After Chinese troops occupied Tibet in 1951 and China's Communist Party first started running the territory, Tibetans were expected to convert to Socialism over time. Land-reform efforts that began in the second half of the 1950s, however, sparked resistance when authorities tried to take land from temples. Anger at these moves helped propel an uprising in 1959, after which the Dalai Lama government fled. With the onset of the Cultural Revolution in 1966, religion came under attack across China. Tibetan monks were forced out of monasteries. Many were forced to violate their vows of celibacy. Communist Red Guards destroyed many temples. The use of the Tibetan language was outlawed. Thousands of Tibetan women silently protested against Communist Chinese occupation and repression in 1959, in Lhasa, Tibet.

These hardline policies began to shift when reformers led by Deng Xiaoping took control of the party after the death of Mao Zedong. Historians view the 1980s as somewhat of a golden era in Beijing's Tibet policy. Monasteries were rebuilt and monks returned. People were again allowed to practice their religion publicly and Tibetan culture was promoted. There were also some tentative talks between the Dalai Lama and the Beijing government, but they broke down without evident progress. After political protests erupted in Lhasa in 1987 and 1989, provoking martial law, the government adopted a hard line against dissent of all kinds in the 1990s.

It was only at the end of the decade that this tough stance was coupled with promises of economic development launched in 2000 to boost the economies of 12 big but poor provinces, including Tibet. Their major goals were to even out inequality and tap into natural resources to support industrial growth back east in the Han Chinese heartland. In 2001, China began construction of what, for Tibet, is the centerpiece of the "Go West" program; the extension of a rail line to connect Lhasa and the rest of the landlocked Tibetan plateau to the booming eastern cities of Beijing, Guangzhou and Shanghai, and their ports. The first trains started running in 2006.

Between 2000 and 2006, there were six rounds of talks between representatives of the Dalai Lama and the Beijing government. The talks made little progress. The Dalai Lama and his government-in-exile, for example, insisted that the Tibet autonomous region be effectively enlarged to include Tibetan areas in nearby provinces as well; a non-starter for the Chinese side. After a meeting in the summer of 2006, negotiations between the Tibetan exiles and Beijing broke off.

7.1.2.3 *Nation/State Level*

A nation/state experiences threats from (1) another country, (2) a domestic non- nation group, (3) an international terrorist, or (4) a natural disaster, all of which are in national security targets. The management of the Fukushima Nuclear Power Plant crisis is also an example of a nation/state level issue, and is analyzed in detail in Section 7.5.2.

7.2 Japanese R&D Ethics

As a Japanese–American, I would like to discuss significant differences in research and development ethics between the US and Japan in this section — through which the reader may be able to understand why the Japanese people are very insensitive to homeland security [4].

7.2.1 *Military vs. Civilian Applications*

Interestingly, Japanese researchers usually hesitate to work on military applications. Japan does not possess any military structure or power. Japan only has a Ministry of Defense in their Constitution since World War II, because the Japanese had been trained to consider the military as ethically bad. Though the current Japanese Constitution was set in 1947 after WWII under the strong influence of the US, it has been unfortunately (or fortunately?) maintained without any significant amendment. In contrast, Americans are proud to voluntarily join the military without any forced conscription. Moreover, because more than 50% of the US's R&D money is created by military-related agents, it is the duty and pride of Americans and researchers based in America to work for military applications, as illustrated in Fig. 7.1. It is totally up to the researcher's ethics to choose between working on, say, a fish-finder development (i.e., a peaceful application) or a submarine sonar development (i.e., a military application), using the same underwater piezoelectric transducer technology (see Figs. 7.1(a) and 7.1(b)).

Figures 7.1(c) and 7.1(d) compare other examples of technologies developed using compact piezo-ultrasonic motors. Seiko Instruments developed a compact helicopter, which can be used as a form of surveillance in battlefields, while my research center ICAT at Penn State developed automatic zoom/focus mechanisms for cellular phone applications in collaboration with Samsung Electromechanics, Korea. However, the same helicopter is also useful for peaceful surveillance (e.g., surveying volcanos), while the world's smallest camera module may also be utilized on a spy camera (as embodied by the Gimbal surveillance system introduced in Section 5.5.2.2). Again, how the product is used depends on the developer's ethics.

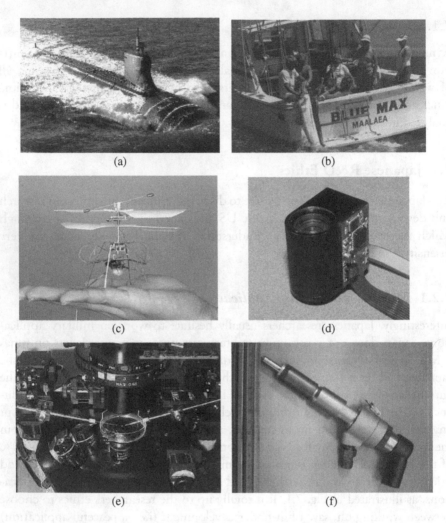

Fig. 7.1.   Examples of America's vs. Japan's R&D foci; reflecting the differences between the US's and Japan's ethics. (a) Sonar on a submarine vs. (b) fish-finder on a fishing boat; (c) surveillance micro-helicopter vs. (d) zoom camera for cellular phones; (e) artificial fertilization equipment vs. (f) a diesel injection valve.

7.2.2 Medical Ethics

Not long ago, I met an enthusiastic, middle-aged Japanese researcher in Tokyo. He was working on a piezoelectric actuator system for artificial insemination/fertilization and proudly declared that the system could impregnate cows precisely and quickly, at the rate of some 200/h (see Fig. 7.1(e)). One question that came to my mind in considering such a system was: What are the moral/ethical implications of this sort of mass production of cows? On the other hand, this equipment could potentially solve the issue of food shortage and starvation in Africa. How one looks at the technology varies across cultures. For example, Americans are interestingly very conservative in their regard toward artificial fertilization technology because of the strong influence of conservative Christianity in America.

Ultrasonic motors have also been utilized for prosthetic arms for handicapped soldiers because they do not produce any audible sound when in operation — a feature well-liked by users of the prosthetic arm(s). Piezoelectric pumps have also come close to being commercialized in the form of artificial hearts. Knowing such advancements are at hand, how much of advancements like these can the reader accept as "ethically acceptable" artificial organs? For example, can the reader accept a prosthetic form of every body-part except the brain?

7.2.3 *Ecological Ethics*

To overcome the "Greenhouse Effect" and other air pollution problems, the Japanese Government formulated a strategy to eliminate the use of diesel engine vehicles, because diesel engines create toxic exhaust gas. In contrast, the German Government adopted a strategy to increase the manufacturing and use of diesel-powered diesel engine vehicles, because diesel is a type of low-grade oil which does not require the consumption of a large amount of electrical energy for distillation. Based on the ethics followed by the Japanese and German governments regarding the use of diesel engine vehicles, which logic will be better accepted? If we calculate the total energy required to wheel both diesel and gasoline from an oil well/rig — though gasoline exhibits better efficiency "from tank to wheel" — it requires more electric energy than diesel "from mine to tank." It is for this reason that, globally, the production of diesel engine vehicles is still expanding drastically. I know this from having consulted for multiple European and Japanese automobile industries regarding piezoelectric diesel injection valve developments (shown in Fig. 7.1(f)), as described in detail in Section 6.5.1.

7.3 **Individual Survival Training**

Some people are rather unprepared for terrorist attacks, natural disasters, or unexpected accidents. Though these incidents may happen to anyone, the final results differ significantly, depending on the individual's level of training for survival. I felt a serious anxiety for the Japanese people's lack of knowledge on individual security during my stay in Japan as a Navy officer. In contrast, during my visit to the Korean Institute of Science & Technology, Korea, in 2011, I experienced a military evacuation drill against a possible North Korean attack. During a meeting held to discuss the Science & Technology (S&T) collaboration possibility between KIST and the US DoD, a loud siren suddenly sounded, interrupting the meeting without any preliminary notice. We were forced to leave the conference room/building immediately to hide in an air-raid shelter with one helmet and a pair of goggles each. These irregular but occasional evacuation drills help keep the people of South Korea prepared in their daily attitudes, for a real military attack.

The unpreparedness of the Japanese for such emergencies is exemplified in the following two incidents. The first happened on February 14, 2013, in which three people were killed and 11 injured after a 21-year-old man allegedly went on a rampage through the tourist district on the tropical island of Guam. The murderer, Chad Ryan De Soto, had allegedly told police that he wanted to hurt as many people as he could, first with his car and then with his knife, according to a declaration filed by the Assistant Attorney General. The second incident, the *In Amenas hostage crisis*, began on January 16, 2013, when al-Qaeda-linked terrorists affiliated with a brigade led by Mokhtar Belmokhtar took over 800 people hostage at the Tigantourine gas facility near In Amenas, Algeria. In that incident, at least 39 foreign hostages were killed along with an Algerian security guard, as were 29 militants. A total of 685 Algerian workers and 107 foreigners were freed. Among the 39 hostages killed were 10 Japanese, 8 Filipinos, 5 Norwegians, 5 British citizens, and 3 Americans [5]. The JGC Corporation (Nikki) disclosed that unfortunately, none of the Japanese workers had been trained on how to handle possible military or terrorist attacks before they started working in the unstable territory in northern Africa, which might have contributed to the high number of Japanese workers killed in the incident.

For contrast, we look at the *Miracle of Kamaishi* that happened during the Great East Japan Earthquake. At a magnitude of 9.0, the Tohoku–Kanto Earthquake triggered a massive tsunami and caused unprecedented damage to a wide area spanning from the Tohoku to Kanto regions. Kamaishi, a city in Iwate Prefecture with a population of approximately 40,000, was hit by a tsunami that topped over 15 m. More than 1,000 people were either killed or went missing. Nearly 30% of the houses were completely or partially destroyed. Despite suffering from this level of damage, almost all of the nearly 3,000 elementary and junior high school students managed to successfully evacuate themselves and stay safe. This amazing fact was acclaimed as the "Miracle of Kamaishi" and drew a large-scale response. *Toshitaka Katada*, a professor of Gunma University, Japan, explained, "The children attribute it to 'Achievements of Kamaishi', a regularly conducted evacuation drill. Their survival is not a miracle considering that they made consistent efforts, did the right things at the right time and obtained the outcome they deserved. Moreover, despite their preparedness, Kamaishi lost five children, so I was somewhat reluctant to use the word 'miracle'." It is important to note that owing to this evacuation drill, even Japanese children can reduce the relatively high death rates typical of similar crisis situations.

In the next section, I introduce a sort of Anti-Terrorism Training program developed by the US Department of Defense. The hypothetical situational background you are to assume to practice responding to the simulated emergency situation(s) is something like the following: You are a Department of Defense civilian officer (like me). In the first part of the exercise, you need to travel without encountering terrorists who try to kidnap or kill you. In the latter part of the exercise, you are a captive/prisoner of enemy forces, and may need to escape from the prison on your own effort.

7.3.1 *Anti-Terrorism Training Program*

I needed to pass several examinations to become a Navy agent working in the ONR Global — Asian Office in Tokyo. One of the important training programs included an "Anti-Terrorism Training" program which consisted of several levels, including a gun/rifle handling course. This section was assembled from my memories of the training program's annual exams. The section provides typical scenarios that could (occasionally) happen when the reader travels around the world. Try to thoroughly consider each scenario in the following eight case studies in Table 7.2 on the next two pages, and answer them before viewing the comments and recommended answers in the following sections.

7.3.1.1 *Active Shooter Situation*

- Do not try to fight an active shooter.
- Hide/protect yourself immediately, without shouting or running.
- In the case of firearms, crouch on the floor, while in the case of grenades, lie flat on the floor.

Recommended Answers: *Situation* 1 = B; *Situation* 2 = B

7.3.1.2 *Air Travel Situation*

- Do not try to confront/fight the suspicious person by yourself.
- Report any incident immediately without hesitation.

Table 7.2.   Anti-terrorism training program.

Active Shooter Situation 1:

In January 2011, six were killed and several wounded in an Active Shooter attack on a political event. In light of recent attacks on Department of Defense personnel like you, you consider how to prepare for the active shooter threat.

What should you do if there is an active shooter incident and evacuation is not possible? *Select the correct response.*

A. Identify items that can be potentially used to attack the active shooter.

B. Seek cover in an area that can be closed off and barricaded.

C. Immediately phone first responders for help.

Air Travel Situation:

While in the airport, be on guard for unattended luggage and keep your luggage with you all times.

A gentleman approached you, and asked you a deal. If you can check one of his bags in your name, he will pay you the $20 fee, and this way he can save the cost of a much higher baggage surcharge. You know that you should not accept the bag. What do you do? *Select the correct response.*

A. Alert airline ticket personnel when you arrive at the counter and give the man's description to security personnel after getting your ticket.

B. Tell the man that checking someone else's luggage is against TSA security policies and his actions could create an incident.

C. Immediately call for security and alert them to the situation.

Active Shooter Situation 2:

One US soldier was killed, and another injured, in a 2009 attack against a recruiting office in Little Rock, Arkansas.

You know you should dive for cover immediately. And you also know that in the event of an attack with firearms you should crouch on the floor. But what should you do in the event of an attack with grenades. *Select the correct response.*

A. Dive for cover and run for the exit at the first possible opportunity.

B. Dive behind something solid and lie flat on the floor.

C. Dive behind something solid and crouch on the floor.

Ground Travel Situation 1:

Items important to vehicle security: internal hood, trunk, gas cap releases, etc. While planning your trip in a major American city, you select a rental car that will not draw attention to you. You also select a car with power door locks and windows to better control access to the car. You also know that a locking gas cap can reduce the chance of tampering.

From a security perspective, what else should you consider? *Select the correct response.*

A. A GPS unit in case you get lost.

B. Air conditioning to allow you to keep windows closed securely in warm weather.

C. Air bags in case of an accident.

Ground Travel Situation 2:

You should choose your routes carefully, especially in strange cities.

You spent the night at your hotel, and the next day you prepare to go to the facility hosting your conference in Air Force Base. TV news reports there is major road construction on the route you planned to take and significant delays are expected.

What is the best way to determine a new route? *Select the correct response.*

A. Use a current city map to plan a new route using major roads.

B. Use a GPS system to plan a new route using side streets.

C. You chose to ask the hotel receptionist for the best route to the facility.

Hotel Security Situation 2:

If someone knocks on your door and you do not know them, do not open the door until you confirm who it is. You ask who it is without opening the door. The person says he is from room service and is bringing complimentary desserts as a welcome to the hotel. Through the peephole, you see a man in a hotel staff uniform holding a tray.

What do you do? *Select the correct response.*

A. Ask him to leave the tray outside the door, listen for him to leave, and then open the door.

B. Use the chain lock, put your foot behind the door, and then open the door slightly to get a better view of the person.

C. Call the front desk to confirm the delivery.

Hotel Security Situation 1:

Hotel rooms should be selected with security in mind. Selecting a room can be important, though you may not have control of your room assignment. However, if you have the choice, consider the following room preferences:

- 3rd to 5th floors – rooms on the 1st and 2nd floors are easily accessible from the outside, and rooms above the 5th floor are difficult to reach by emergency services.
- Not immediately adjacent to fire escapes or emergency exits – criminals may target these rooms due to the ease of escape.
- No balcony – criminals may use balconies to enter rooms and to go from one room to the next.

Hotel Security Situation 3:

Hotel room invasions have been increasing in the United States. You are traveling and spending a night in a hotel. Suddenly, you hear a crash and the door and two armed men wearing hoods enter your room.

Waving their guns, they shout for you to get on the floor and be quiet. How do you respond? *Select the correct response.*

A. Do not resist, slowly crouch to the floor, and do not make any sudden movements that might be considered threatening.

B. Attempt to seize a weapon from one of the assailants if one is close to you.

C. Dive for cover behind a solid piece of furniture and then run for the door.

- Do not wear noticeable/prominent costumes, including package-tour uniforms or badges/identifying tags, or display unnecessary group behavior (unless it is regulated). This increases the risk of being targeted by the attacker.

Recommended Answer: C

The general response of an average Japanese citizen seems to be A. However, adopting this approach may occasionally mean only acting when it is too late; the suspicious person could have already escaped, or the bomb might explode before action A is carried out. B is the riskiest choice for a civilian.

7.3.1.3 *Ground Travel Situation*

- Consider the traveler's security from terror attacks as the highest priority.
- When you rent a car, check for any suspicious items (bombs, wiretaps, etc.) inside the car, hood and trunk.
- Complete door and window closing functions are essential.

Recommended Answer: Situation 1 = B

- A GPS system is just for your convenience and is not related to your security. Airbags are a necessity for your personal safety. Note that a detour can occasionally be a trap set by a terrorist group.
- Do not choose side/local roads for your drive.
- Hotel persons may also be bribed by the terrorist group. The average Japanese citizen may easily believe persons wearing the hotel uniform, and not consider that it may be a dangerous situation in foreign countries, because a similar situation in Japanese society is rather safe.

Recommended Answer: Situation 2 = A

7.3.1.4 *Hotel Security Situation*

Regular tourists prefer rooms on higher floors, facing a busy street; the most prohibited attitude for a Navy officer. Select a hotel room:

- *On the 3rd to 5th floors* — rooms on the 1st and 2nd floors are easily accessible from the outside, and rooms above the 5th floor are difficult for emergency services to reach. The sacrifice of a good view from a higher floor is necessary in order to protect yourself from surveillance. Facing a busy street helps kidnappers easily monitor your behavior in the hotel room.

- *Not immediately adjacent to fire escapes or emergency exits* — criminals may target these rooms due to the ease of escape.
- *Without a balcony* — criminals may use balconies to enter rooms and to go from one room to the next.

Be cautious when opening the room door. If intruded upon, do not resist the intruder; slowly crouch to the floor, and do not make any sudden movements that might be considered threatening.

Recommended Answer: Situation 2 = A; *Situation* 3 = A.

7.3.2 *Kidnap Escaping Program*

If you are kidnapped, in the worst scenario, you may need to give serious consideration to how you can escape from the place you are being held. While you may expect a rescue by a *SWAT* (Special Weapons and Tactics) team in the US, no such rescue effort can be expected from Japanese forces for Japanese citizens, for example, when the kidnapping occurs outside of Japan — as seen in the *In Amenas hostage crisis* in 2013, when the Japanese government did not send a military rescue team to Algeria, even though more than ten Japanese engineers were kidnapped. Note that according to the global security rule, ransom money should not be paid to terrorists in exchange for kidnapping victims.

Without any military training, it may be difficult for the reader to take the actions described in the following segments; but theoretical knowledge of the basic idea is better than nothing. Never consider taking any suicidal actions, such as engaging in aggressive resistance, against terrorists. The basic plan is to *return with honor*. Try answering the questions in Table 7.3 (viewable on the next two pages) prior to reading the following sections for the answers.

7.3.2.1 *Opportunities to Escape*

In any situation of captivity and/or when you are being transported, you may be able to identify various opportunities to escape.

- *Initial capture*

This phase may offer the best opportunity to escape. Your captors would not have had much experience holding or restraining you, and you may be in your best physical condition and still be in possession of the clothing and equipment needed for survival and evasion.

Table 7.3.   Kidnap escaping program.

Kidnap Escape Situation 1:

Suppose that you are in "In Amenas" hostage crisis, Algeria. You are in your private apartment, as illustrated below. You heard a machine gun sound in the next door. Then, you heard a terrorist voice outside your apartment door, **"Open the door!"**. How will you respond to this? *Select the correct response.*

A. Following the order, you will open the door.
B. You will hide behind the cupboard.
C. You will fall face down on the floor below the bed.

B		
Cupboard		TV
Bed		Sofa
	C	A Door

Kidnap Escape Situation 2:

You escaped from the gas facility located south-west of In Amenas, when al-Qaeda-linked terrorists attacked the facility. You try to find the best place for an evasion shelter. Where do you choose? *Select the correct response.*

A. River-side cliff for drinking water
B. Flat hill to be found easily from a helicopter of your rescue team
C. Deep forest

Kidnap Escape Situation 3:

You escaped from the gas facility located south-west of In Amenas, when al-Qaeda-linked terrorists attacked the facility. You hid in a deep forest by making an evasion shelter. However, you need to return to the safe US-related base camp, to which you do not know the direction. The next day you see an older man approaching to your shelter, and spotted you. What will you do? This older man does not wear a military uniform without weapons, and seems to be a local resident. *Select the correct response.*

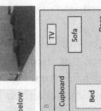

A. Run
B. Prepare Pointee-Talkee*
C. Shoot the older man

*A language aid containing selected phrases in English opposite a translation in a foreign language. It is used by pointing to appropriate phrases.

Kidnap Escape Situation 4:

You were found unfortunately by the chasers, and captured. The enemy asked you "Who are you and what are you doing in my country?". How do you respond? *Select the correct response.*

A. Quickly respond with full name and true mission statement
B. Remain silent
C. Stall, provide first name and state that you were doing nothing wrong

Kidnap Escape Situation 5:

You were captured unfortunately and sent to the internment. You hear knocking on the wall. A fellow captive is communicating with you in secret using the quadratic alphabet, or tap code. Review the standard quadratic alphabet chart to see how captives can slowly spell out words through tapping, knocking, coughing, blinking, etc. in sequences that from letters of the alphabet.

Then the sound of your fellow captive transmitting a word to you in code seems to be as follows. The longer silences indicate a new letter being coded to you. *Select the word being transmitted from the list.*

Talking loudly

Talking through the wall

•••••• •• ••••• ••••• •••

A. JPRA
B. WITS
C. WILL

Kidnap Escape Situation 7:

During the captive period, you have been ordered to record the terrorist's message about your capture, by reading the paper given by the captor. Which of the following statements should you deliver during recording? *Select all that apply from the list.*

A. Explain the innocent circumstances leading to your capture
B. Describe sustained injuries
C. Identify Senior Sergeant with you in captivity
D. Apologize for US presence in the region and say that US forces should withdraw
E. State that you are being treated well by your captors and express regret for loss of innocent lives

Kidnap Escape Situation 6:

During the extraction phase the interrogator may attempt various interrogation methods to gain the desired information. A solid understanding of the principal methods used in interrogation will greatly assist you in resisting exploitation. What do you categorize the following method? *Select the correct response.*

This is the "bad cop/good cop" strategy. The captor has one individual physically abuse and/or threaten you, while a "friendly" captor "saves" or "protects" you. The captor's goal is to elicit cooperation through your gratitude toward the "good cop". Remember, both are interrogators and the "good cop" set this up. Don't be grateful.

A. Disgrace
B. Accusation
C. Fear and despair
D. Threat and rescue
E. Shock and surprise
F. Friendly
G. Non-interrogator
H. Technical methods

Kidnap Escape Situation 8:

You have been asked to participate in an interview with local media. Which of the following statements should you use during the course of the conversation? *Select all that apply from the list.*

A. "I would like to thank my captors for their good treatment"
B. "For the most part, our group is doing well"
C. "Our enemies are not responsible for the actions declared by US forces"
D. "I would like to return to US Government control"
E. "I would like to speak to the US Embassy"

- *Movement*

Look for opportunities to escape during movement. Whether on foot or in a vehicle, you may be able to break contact with your captor and get into good evasion cover. Holding facilities during transit may also be low in security and not designed for prisoner detention.

- *Internment*

In an internment facility, the challenge to escape increases with each passing moment. Your physical condition will likely worsen, and you will likely be held in a more secure facility which may isolate you from friendly forces. It is still possible to escape, but it is typically more challenging than during initial capture and movement.

Recommended Answer: Kidnap Escape Situation 1 = C

This is the actual behavior of the persons who survived in the "In Amenas hostage crisis" in 2013. They never opened the door. Even when they were afraid, they did not obey the terrorists. The Japanese workers who perished might have obeyed the terrorists' orders to "Open the door" out of fear.

7.3.2.2 *Evasion Shelter*

An evasion shelter must first satisfy the criteria for a hole-up site. Consider using or modifying an existing natural shelter that already provides protection from the environment such as:

- In rural areas: fallen trees, rock crevices, trees with heavily laden boughs close to the ground, etc.
- In an urban environment: areas void of people, bombed-out buildings or structures, dumps, culverts, etc.

If necessary, improve a natural shelter using:

- Natural materials: tree boughs, slabs or tree bark, rocks, fallen timber, grasses, leaves or other vegetation, blocks of snow, etc.
- Manmade materials: poncho, garbage, scrap wood, cardboard, plastic bags, etc.

Evasion shelters should comply with the *BLISS principle*.

- *Blend*
- *Low*
- *Irregular*
- *Small*
- *Secluded*

Recommended Answer: Kidnap Escape Situation 2 = C

Your tracer team will first come to the river-side. A flat hill is important for the final helicopter meeting place, but dangerous for occupying for a long period.

Recommended Answer: Kidnap Escape Situation 3 = B

You need to find a way to return to the safe US base related area. Thus, you should have the courage to politely ask a local resident for directions. The *pointee-talkee* is a mandatory tool in an unfamiliar country. Though you may use an electronic "Talking Dictionary" in regular life, never consider using this machine in a rural area, because after a couple of weeks have passed the battery would already have gone flat!

7.3.2.3 *Captive Place*

Unfortunately, you are captured. You should learn how to behave in the place of captivity without losing dignity.

- Remember how much you can disclose to the terrorists.
- Any possible communication with other captives is essential.
- Keep your dignity and do not lose the courage to escape, nor the hope of rescue by the US SWAT team.

Recommended Answer: Kidnap Escape Situation 4 = C

Do not respond to the enemy's questions quickly, instead stall all the time. Remember: *answer minimally* (the less information you reveal, the better). However, remaining silent is very risky. State that *you were doing nothing wrong* to keep your dignity.

You should collect as much information in the internment as possible. Communication with other captives can be accomplished in many ways: talking loudly, talking through the wall, Morse code, etc., via the English language. Taps are an example of a means used to identify letters. Figure 7.2 shows the *Quadratic Alphabet* used in the US military. In this system, the first series of taps denotes the row. After a short pause, the second series of taps denotes the column. The desired letter will be in the block where the row and column meet. For example, three taps designate the third row (letter L), followed by a pause, and four taps to designate the letter O. Numbers are sent by a slow tapping until the number desired is reached. This *Tap Code* is better than Morse when short and long sounds cannot be distinguished much, as in the case of wall tapping. The versatility of the Tap Code is almost unlimited. The Code can be transmitted via instruments, shadows, coughing, blinking, etc.

1st Series of taps Id's the = ROW
2nd Series of taps Id's the = COLUMN

Fig. 7.2.   Quadratic alphabet or tap code in the US military.

Recommended Answer: Kidnap Escape Situation 5 = B

Remembering the Tap Code Chart is easy. Just notice that C and K are over-lapped on the first row.

7.3.2.4 *Interrogation Methods*

During the extraction phase, the interrogator may attempt various interrogation methods to gain the desired information. A solid understanding of the principal methods used in interrogation will greatly assist you in resisting exploitation.

Recommended Answer: Kidnap Escape Situation 6 = D

The captor often insists that you confess that you did wrong, and that US forces should withdraw. You may mention:

- Your name
- Your physical condition
- Other captives with you

However, never mention:

- That you did wrong
- That you were treated well

Recommended Answer: Kidnap Escape Situation 7 = A, B, C
Recommended Answer: Kidnap Escape Situation 8 = B, D, E

Though declaring that "I would like to speak to the US Embassy" is a regular right during normal country vs. country wars, international terrorists may not accept this request.

7.4 Non-Nation Group: Crisis Management

A non-nation group, such as a particular minority race in a nation, may experience sudden and unfavorable changes in their environment, such as the suppression of activities from the government, or a threat by another competitive race. Companies too are non-nation groups that may experience similar unexpected situations. For example, when a sudden change in the economic environment is triggered by a major event such as an "oil crisis"; when a new aggressive rival in your industry appears without warning; or when a major client suddenly cancels their purchase of your products, which may cause a crisis in your company. Because they are *unexpected situations*, how these situations should be handled differ from how one handles normal situations. The key to effective risk/crisis management is *quick decision-making with insufficient information*. Therefore, do not delay the decision by wasting time in collecting complete information for the perfect solution.

7.4.1 *Civil War*

Though it is considered a form of terrorism from the viewpoint of the Chinese government, the protest by the Tibetan monks in the Drepung and Sera monasteries can be perceived as a civil war for independence from the other point of view. A similar situation can be found in the Xinjiang Uyghur Autonomous Region, the largest administrative division in the northwest region of China, which contains the disputed territory of Aksai Chin administered by China. These disputes are basically considered an oppression of the Chinese government on the Buddhist and Muslims in those areas. Human rights organizations have been critical of the Beijing government's approach to human rights in those regions when cracking down on separatist convulsions. In these cases, the compromised solution is made by balancing the "military power" from the Chinese government and "humanitarian" pressure from the international community.

Most recent civil wars were triggered by the pursuit of *natural resources*, such as oil, natural gas, diamonds, and gold. Since these natural resources are geologically unevenly distributed, anti-government parties try to possess these resources and initiate civil war. An example of this is the Sierra Leone Civil War, which began in March 1991 when the Revolutionary United Front (RUF), with support from the special forces of Charles Taylor's National Patriotic Front of Liberia (NPFL), intervened in Sierra Leone in an attempt to overthrow the *Joseph Momoh* government [6]. During the first year of the war, the RUF took control of large swathes of territory in eastern and southern Sierra Leone, which were rich in alluvial diamonds. By the end of 1993, the Sierra Leone Army (SLA) had succeeded in pushing the RUF rebels back to the Liberian border; however, the RUF managed to recover and fighting continued. In May 1997, a group of disgruntled SLA officers staged a coup and established the Armed Forces Revolutionary Council (AFRC) as the new government of Sierra Leone. In January 1999, world leaders intervened diplomatically to promote negotiations between the RUF and the government. As a result, the Lome (capital city) Peace Accord was signed in March 1999. It gave *Foday Sankoh*, the commander of the RUF, the vice presidency and control of Sierra Leone's diamond mines in return for a cessation of the fighting and the *deployment* of a UN peacekeeping force to monitor the disarmament process. Unfortunately, RUF's compliance with the disarmament process was inconsistent and sluggish, and by May 2000, the rebels were advancing again on Freetown. It was then that the United Kingdom declared its intention to intervene in the former colony in an attempt to support the weak government of President *Ahmad Tejan Kabbah*. With help from a renewed UN mandate and Guinean air support, the British Operation Palliser finally defeated the RUF, and took control of Freetown. In 2002, President Kabbah declared the Sierra Leone Civil War, over. In this case, the initial solution had been reached by balancing the "military power" from the government and the rebels' "economic" attraction to diamonds.

7.4.2 *Company Risk Management*

Strategic decision-making is essential even in a company for promoting a project. There are three types of decision-making mistakes:

- Mistake in decision — a mistake in the strategic choice selected.
- Mistake in procedure — mistakes in the handling process, though the strategy selected is appropriate.
- Mistake in handling timing — incorporating the appropriately strategy and procedure too slowly and/or too late.

7.4.2.1 *Decision Tree*

Let me introduce the concept of the *"Decision Tree"* to help you understand strategic decision-making. For the case study, we consider an Oil Mining Company's

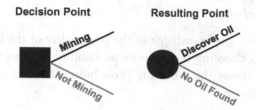

Fig. 7.3.   Decision-making point and resulting point in the decision tree map.

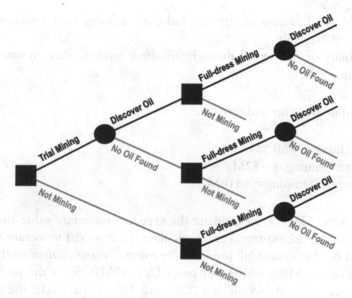

Fig. 7.4.   Decision tree for oil mining.

decision-making in a simple example. There are two different key points in the decision tree: the Decision-Making point and Resulting point. As illustrated in Fig. 7.3, the decision to mine or not to mine (shown as "Mining" or "Not Mining" in Fig. 7.3) should be decided at the decision-making points, while discovering oil or not finding oil (shown as "Discover Oil" and "No Oil Found" in Fig. 7.4) are the results at each resulting point. While the oil mining business receives huge profits once oil is actually discovered, it possesses high risks in cases where no oil can be found. Furthermore, mining costs are really expensive. Therefore, the manager needs to make multiple decisions during the business process. Figure 7.4 shows the decision tree for the oil mining business, with the available options being: (1) trial mining and/or (2) full-dress mining.

7.4.2.2 *Risk Analysis*

Risk Analysis requires the knowledge of the probability of the incident. Though slight variations in choosing the accurate probability changes the final decision dramatically, we assume the following probabilities for our case study for the sake of further analysis:

- Probability of finding oil through trial mining = 60%;
- Probability of finding oil through immediately engaging in full-dress mining = 60%;
- Probability of finding oil through full-dress mining, after a successful trial mine = 90%; while
- Probability of finding oil through full-dress mining, after an unsuccessful trial mine = 10%.

Regarding the costs and revenue:

- Trial mining = – $0.5M.
- Full-dress mining = – $2M.
- Successful oil mining = +$10M.

Now using Fig. 7.5, we estimate the expected monetary value analysis for this oil mining decision-tree. Let us calculate the expected monetary value for each point on the A route (all top lines) by using a *reverse analysis* method. As a result, we obtain $10M at the end-point (A (+10M)). Since the probabilities for "discover oil" and "no oil" are 90% and 10% respectively, the expected value calculated at the prior step should be $9M. Similarly, since full-dress mining costs $2M, the expected value two steps prior should be $7M. Again, since the probabilities for "discover" and "no oil" are 60% and 40% respectively, for trial mining, the expected value three steps prior should be $4.2M. Then, by subtracting the trial mining cost ($0.5M), the initial expected monetary value can be calculated as $3.7M. We should at least choose the cases where the initial expected monetary value should exceed the total required cost, i.e., only A and G remain. In this scenario, because of the relatively high probability (60%) of "discover[ing] oil," skipping the trial mining creates a larger profit, i.e., G is the best choice. However, if this probability were to be 30%, which scenario do we choose? Similar to the above calculation, for the A route, $7M × 0.3 – $0.5M = $1.6M, while for the G route, $10M × 0.3 – $2M = $1.0M. In this scenario, the A route is the best choice, i.e., trial mining is really necessary.

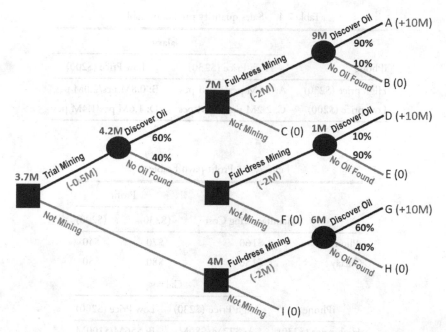

Fig. 7.5.   Expected monetary value analysis for oil mining.

7.4.2.3 *Game Theory*

Let us recall the game theory learned in Section 3.3. We take an example based on the battle of the Apple iPhone and Samsung Galaxy product lines, which are the most competitive mobile phone products in the world. Because the price set changes the potential market share and revenue earned by each product significantly, both Apple and Samsung Electronics are very sensitive in regard to this. To make the situation simple, we consider the following assumptions:

- How to set the sales price: Only two prices levels — high price ($230) and low price ($200).
- Market: Lower price generates larger sales quantity.
- Profit = Sales price — Manufacturing cost.
- Profit rate: Lower price makes lower profitability.

STEP 1: Create the sales quantity prediction table

Table 7.4 shows a prediction table of the sales quantities for the iPhone and the Galaxy according to the high or low setting prices. We assume that the iPhone

Table 7.4.   Sales quantity prediction table.

iPhone	Galaxy	
	High Price ($230)	Low Price ($200)
High price ($230)	A: 1.1M pcs/1.0M pcs	B: 0.8M pcs/2.0M pcs
Low price ($200)	C: 2.2M pcs/0.7M pcs	D: 1.6M pcs/1.4M pcs

Table 7.5.   Profit payoff table.

	Manufacturing Cost	Profit	
		($230)	($200)
iPhone	$160	$70	$40
Galaxy	$150	$80	$50

iPhone	Galaxy	
	High Price ($230)	Low Price ($200)
High price ($230)	A: $77M/$80M	B: $56M/$100M
Low price ($200)	C: $80M/$56M	D: $64M/$70M

has slightly higher *Brand Name Value* than the Galaxy, which creates a slight difference (1.1 million pieces vs. 1.0 million pieces) when sold at the same high price ($230) in Cell A.

STEP 2: Create the profit payoff table

Using the profit calculation chart in the top-half of Table 7.5, we create the profit payoff table. The manufacturing cost for one Galaxy is $10 lower than an iPhone, which reflects on the profit for both high and low sales price cases. If you multiply the profit with the production quantity in Table 7.4, you will get the profit payoff table as given in the bottom-half of Table 7.5.

STEP 3: Strategy decision-making

If the two parties (Apple and Samsung Electronics) could come to an agreement in collusion with one another, it is likely that both would choose the "High price" option (Cell A), leading to the $77M/$80M profit for both. This is politically possible, like what we see in OPEC on the international level.

If an agreement is not allowed (e.g., in a free market, collusion may be illegal in one nation), Apple will use the following strategies: (a) If Samsung takes the "High price" option, it is better for Apple to take the "Low price" option (Cell C) because the profit will increase from $77 M (Cell A) to $80M per year. (b) If Samsung sets a "Low price," it is better for Apple to also set a "Low price" (Cell D), because the profit will be $64M — higher than the $56M in Cell B. Vice versa, Samsung Electronics will consider the following strategies: (a) when Apple chooses to set a "High price," it is better for Samsung to set a "Low price" (Cell B) because the profit will increase from $80M (Cell A) to $100M per year. (b) When Apple chooses to set a "Low price," it is better for Samsung to also set a "Low price" (Cell D), because the profit will also be $70M — higher than the $56 M projected in Cell C. We therefore conclude that Cell D is the *dominant strategy* for both Apple and Samsung; a move called the *Nash Equilibrium* solution in this Prisoner's Dilemma's case.

7.4.3 *Military FM 3-0 Operations Manual*

The sentences in the subsections below with item numbers (e.g., "5-2" or "5-3" in Section 7.4.3.3) are extracts from the *Field Manual, FM* 3-0 [7].

7.4.3.1 *FM 3-0 Operations Principle*

Former US Army General, William S. Wallace mentioned the following in the preface of FM 3-0 Operations:

> *America is at war, and we live in a world where global terrorism and extremist ideologies are realities. The Army has analytically looked at the future, and we believe our Nation will continue to be engaged in an era of "persistent conflict" — a period of protracted confrontation among states, non-state, and individual actors increasingly willing to use violence to achieve their political and ideological ends. The operational environment in which this persistent conflict will be waged will be complex, multi-dimensional, and increasingly fought "among the people."*

FM 3-0 describes an operational concept where *commanders employ offensive, defensive, and stability or civil support operations simultaneously as part of an interdependent joint force to seize, retain, and exploit the initiative, accepting prudent risk to create opportunities to achieve decisive results*. This edition will take us into the urban battlefields of the 21st century among the people, without losing our capabilities to dominate the higher conventional end of the spectrum of conflict. It recognizes that *we will achieve victory in this changed environment of persistent conflict only by conducting military operations in concert with diplomatic, informational*, and *economic efforts*. Success in the battlefield alone is no longer enough; the final victory requires concurrent stability operations to lay the foundation for lasting peace.

Although the strategic environment and operational concepts have changed, *soldiers remain the centerpiece and foundation of the Army*. These soldiers are led by leaders who are proficient in their core competencies, who possess sufficient and broad enough a knowledge to adapt to conditions across the spectrum of conflict, and who are courageous enough to face vulnerabilities among enemies and exploit opportunities in the challenges and complexities of our operating environments. As leaders, it is our obligation to understand and be proficient at employing soldiers in full spectrum operations.

7.4.3.2 *Crisis/Risk Management Flowchart*

Figure 7.6 shows a basic flowchart for crisis/risk management. Taking the Nuclear Power Plant accident or the oil mining project as examples, the following actions should be considered:

- **Crisis/risk Occurrence**
- **Initial Stage:**

   o   Monitor the crisis.
   o   Setup a command system and temporary budget.

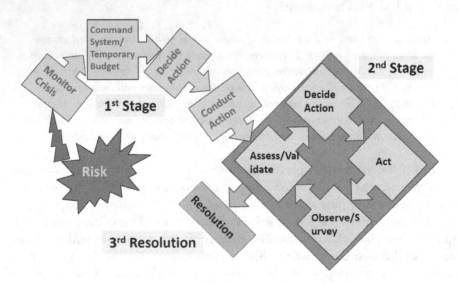

Fig. 7.6.   Basic flowchart for crisis/risk management.

- o Decide on the first action.
- o Conduct the action.

- **Second Stage:**

  - o *Assess/validate* the conducted action.
  - o *Decide* on the second action.
  - o *Conduct* the action.
  - o *Observe/survey* the result.
  - o *Assess/validate* the second action's result.
  - o [Repeat the above routine as many times as required].

- Resolution

  "Decide → Conduct → Observe → Validate" is a cycle of the action decision.

7.4.3.3 *Command System*

A Democratic *leader* is sometimes respected in a normal situation, but a strong *commander* is highly required in a crisis situation. Items number 5-XX in the following segment were extracted from FM 3-0.

5-2. The commander is the focus of command and control. Through it, commanders assess the situation, make decisions, and direct actions. However, *commanders cannot exercise command and control alone* except in the smallest organizations. Thus, commanders perform these functions through a command and control system — the arrangement of personnel, information management, procedures, and equipment and facilities essential for the commander to conduct an operation. An effective command and control system is essential for commanders to conduct (plan, prepare, execute, and assess) operations that accomplish the mission decisively.

5-3. *Soldiers* are the most important element of a command and control system. Their actions and responses are a means of controlling operations. Soldiers also assist commanders and exercise control on their behalf. *Staffs* perform many functions that help commanders exercise command and control. These include:

- Providing relevant *information and analysis*;
- Maintaining running *estimates and making recommendations*;
- Preparing plans and orders;
- Monitoring operations;
- Controlling operations;
- Assessing the progress of operations.

7.4.3.4 *Principles of Operations*

The *nine principles of war* represent the most important non-physical factors that affect the conduct of operations at the strategic, operational, and tactical levels. Though the Army published its original principles of war after World War I, they are still valid in crisis and corporate risk management.

1. Objective — Direct every operation towards *a clearly defined, decisive, and attainable objective*.
2. Offensive — Seize, retain, and *exploit the initiative*.
3. Mass — Concentrate the effects of combat power at the decisive place and time.
4. Economy of force — Allocate minimum essential combat power to secondary efforts.
5. Maneuver — Place the enemy in a disadvantageous position through the flexible application of combat power.
6. Unity of command — For every objective, ensure unity of effort under *one responsible commander*.
7. Security — Never permit the enemy to acquire an unexpected advantage.
8. Surprise — Strike the enemy at a time or place or in a manner for which he is unprepared.
9. Simplicity — Prepare clear, uncomplicated plans and clear, *concise orders* to ensure thorough understanding.

7.4.3.5 *Fog of Uncertainty*

Under an uncertain situation where any decision could bear high risks (i.e., a *Fog of Uncertainty*), the commander needs to plan necessary actions very quickly in order to escape from ruin or destruction.

5-8. In battle, commanders face a thinking and adaptive enemy. Commanders estimate, but *cannot predict*, the enemy's actions and the course of future events. Two key concepts for exercising command and control in operations are *battle command* and *mission command*. Battle command describes the commander's role in the operations process. Mission command is the Army's preferred means of battle command.

5-9. Battle command is the art and science of understanding, visualizing, describing, directing, leading, and assessing forces to impose the commander's will on a hostile, thinking, and adaptive enemy. Battle command applies leadership to translate decisions into actions to accomplish missions. Battle command is guided by professional judgment gained from *experience, knowledge, education, intelligence,* and *intuition,* which is driven by commanders.

5-10. Successful battle command demands *timely and effective decisions* based on applying judgment to *available information.* It requires knowing both when and what to decide. It also requires commanders to evaluate the quality of information and knowledge. Commanders identify important information requirements and focus subordinates and the staff on answering them. Commanders are aware that, once executed, the effects of their decisions are frequently irreversible. Therefore, they anticipate actions that follow their decisions.

5-11. Commanders continuously combine analytic and intuitive approaches to decision-making to exercise battle command. *Analytic decision-making* approaches a problem systematically. The analytic approach aims to produce the optimal solution to a problem from among the solutions identified. The Army's analytic approach is the military decision-making process (*MDMP*). In contrast, *intuitive decision-making* is the act of reaching a conclusion that emphasizes on pattern recognition based on knowledge, judgment, experience, education, intelligence, boldness, perception, and character. This approach focuses on the assessment of the situation vs. making a comparison of multiple options (FM 6-0). It relies on the experienced commander's and staff member's intuitive ability to recognize the key elements and implications of a particular problem or situation, reject the impractical, and select an adequate solution.

5-12. The two approaches are not mutually exclusive. Commanders may make an intuitive decision based on situational understanding gained during the MDMP. If time permits, the staff may use a specific MDMP step, such as war-gaming, to validate or refine the commander's intuitive decision. When conducting the MDMP in a time-constrained environment, many techniques rely heavily on intuitive decisions. Even in the most rigorous, analytic decision-making processes, intuition sets boundaries for analysis.

5-13. Commanders understand, visualize, describe, direct, lead, and assess throughout the operations process. First, they develop a personal and in-depth understanding of the enemy and operational environment. Then they visualize the desired end state and a broad concept of how to shape the current conditions into the end state. Commanders describe their visualization through the commander's intent, planning guidance, and concept of operations in a way that brings clarity to an uncertain situation. They also express gaps in relevant information, as required in the commander's critical information requirements (CCIRs). Direction is implicit in command; commanders direct actions to achieve results and lead forces to mission accomplishment.

5-14. Effective battle command requires commanders to continuously assess and lead. Assessment helps commanders better understand current conditions and broadly describe future conditions that define success. They identify the difference between the two and visualize a sequence of actions to link them. Commanders lead by force of example and personal presence. Leadership inspires Soldiers to accomplish things that they would otherwise avoid. This often requires risk. Commanders anticipate and accept risk to create opportunities to seize, retain, and exploit the initiative and achieve decisive results.

5-15. Battle command encourages the leadership and initiative of subordinates through mission command. Commanders accept setbacks that stem from the initiative of subordinates. They understand that land warfare is chaotic and unpredictable and that action is preferable to passivity. They encourage subordinates to accept calculated risks to create opportunities, while providing intent and control that allow for latitude and discretion.

The key points may summarized as follows:

- There should only be *one responsible commander* (no democracy!) for crisis management.
- Combine *analytic and intuitive approaches* to decision-making.
- *Speed* is the highest priority in decision-making even under insufficient information.
- Transfer the initiative/decision responsibility to as low a subordinate as possible.
- Encourage subordinates to accept *calculated risks*. Commander accepts set-backs of the initiatives by subordinates.

An economist from the University of Chicago, *Frank Knight* (1885–1972), distinguished between risk and uncertainty in his work "Risk, Uncertainty, and Profit". He defined *Risk* as the threat from which the probability of its occurrence can be estimated, while *Uncertainty* was defined as the threat from which the probability of its occurrence cannot be estimated [8]. He also distinguished three uncertain incidents: *Mathematical Probability* (the probability estimated by a mathematical analysis, like the probability of the number that appears when the roll of dice is considered to be a random event); *Statistical Probability* (the probability estimated from the statistical data such as the mortality rate or the remaining lifetime for a certain age); and *Supposition* (a one-time incident for which it is difficult to calculate the probability of its occurrence).

In the case of the previous oil mining strategic project, we used the following Statistical Probabilities for calculating the expected values:

- Probability of finding oil through trial mining = 60%.
- Probability of finding oil through immediate full-dress mining = 60%.

- Probability of finding oil through full-dress mining, after a successful trial mine = 90%, while the probability of finding oil through full-dress mining, after an unsuccessful trial mine = 10%.

If the probability is difficult to estimate, i.e., an "uncertainty," we may assume a 50–50% chance of the occurrence.

7.5 National Crisis Management

Crises can be categorized into five types: (1) Natural Disasters, (2) Infectious Diseases, (3) Enormous (Large-Scale) Accidents, (4) Terrorist/Criminal Incidents, and (5) Wars/Territorial Invasions. In this segment, we consider issues requiring national-level crisis management; in particular, issues pertaining to international terrorism, natural disasters, and large-scale accidents such as a nuclear power plant meltdown.

7.5.1 *International Terrorism*

In addition to civil wars (discussed in the previous section), international terrorism has gained more attention particularly after the 9/11 incident (also known as the World Trade Center attack).

7.5.1.1 *Reaction to International Terrorism*

International terrorist groups such as *al-Qaeda* (an Islamist organization founded by *Osama bin Laden* and *Abdullah Azzam*, who are global militants) create global networks and conduct terroristic actions internationally [9]. The foundation originated from the Soviet war in Afghanistan. It operates as a network comprising both a multinational, stateless army and a terrorist group which may consist of Islamists, extremists, Wahabists and jihadists. It has been designated as a terrorist organization by the United Nations Security Council, the North Atlantic Treaty Organization (NATO), the European Union, the United States, Russia, India, and various other countries. Al-Qaeda has carried out many attacks on targets it considers *kafir* (disbeliever, infidel); both on civilians as well as military entities via *suicide attacks*. During the Syrian civil war, al-Qaeda factions started fighting each other, as well as the Kurds and the Syrian government. The US government responded to the September 11 attacks by launching the *War on Terror*. With the loss of key leaders and culminating in the death of *Osama bin Laden*, al-Qaeda's operations have devolved from actions that were controlled from the top down, to actions by franchise associated groups and lone-wolf operators.

The key strategy against these international terrorist groups is composed of (1) domestic and (2) international actions. Domestically, the US established the *Department of Homeland Security* (DHS) in 2002 after the September 11 attack. The Department of Homeland Security has a vital mission: "to secure the nation from the many threats we face." While it largely focuses on federal preparations to deal with terrorism, the DHS also manages other duties including border security, cyber security, customs and emergency management. The *Associated Press* reported on September 5, 2007, that DHS had scrapped an anti-terrorism data mining tool called *ADVISE* (Analysis, Dissemination, Visualization, Insight, and Semantic Enhancement) after the agency's internal Inspector General found that pilot testing of the system had been performed using data on real people without the required privacy safeguards in place. The system, in development at Lawrence Livermore and Pacific Northwest National Laboratory since 2003, has cost the agency $42 million to date. This is not a new controversy over the ADVISE system. In fact, in March 2007, the Government Accountability Office stated that "the ADVISE tool could misidentify or erroneously associate an individual with undesirable activity such as fraud, crime or terrorism." Homeland Security's Inspector General later said that ADVISE was a poorly-planned and a time-consuming data mining tool for analysts to use, and that it also lacked adequate justifications [10].

An essential international action is to establish the common global norm to prohibit terroristic actions, i.e., to create a global regime on anti-terrorism to prevent and/or provide countermeasures. After the September 11 incident, the United Nations passed an anti-terrorism resolution (Resolution 1373), and set up the Counter-Terrorism Committee in 2001 to collect internationally, information on groups and individuals associated with/related to al-Qaeda and on money laundering. The Asia-Pacific Economic Cooperation (APEC), ASEAN Regional Forum (ARF), and Shanghai Cooperation Organization (SCO) are also acting on the anti- terrorist campaign, in parallel to the World Bank and IMF, which regulate money laundering in order to terminate the flow of (monetary) resources into terrorist groups. Money laundering is the process of transforming the proceeds of crime into ostensibly legitimate money or other assets. However, most anti-money laundering laws openly conflate money laundering (which is concerned with the source of funds) with terrorism financing (which is concerned with destination of funds) when regulating the financial system. Another issue is to prohibit the *transporting* of weapons of mass destruction (WMDs) to terrorist groups. This can be achieved by implementing the "Container Security Initiative" started in the US for accurately checking the contents of containers transported by ships, and the *Proliferation Security Initiative* (PSI). The Container Security Initiative and PSI are examples of global efforts to stop the trafficking of WMDs, their delivery systems, and related materials to and from state and non-state actors of proliferation concern. Launched by former United States President George W. Bush in May 2003 at a meeting in Kraków, Poland, PSI has now grown to include the endorsement of 103 nations around the world, including Russia, Canada, the United Kingdom, Australia, France, Germany, Italy, Japan, the Netherlands, Poland, Singapore, New Zealand, Republic of Korea, and Norway. Despite possessing overwhelming support from nearly half of the Members of the United Nations, a number of countries have expressed opposition to the initiative, including India, China, and Indonesia.

7.5.1.2 *Enforcement of Anti-Terrorism Regime and Anti-Terrorism War*

There is an empirical rule which follows dealings with terrorists; that once we start negotiating with the outrageous requirements of a terrorist group prior to a terrorist attack, we can expect to face further severe attacks from the terrorist group. We can therefore conclude that the relationship between the international terrorist groups and our nations is merely a "zero-sum" game, without any compromises — which makes establishing a global anti-terrorism regime extremely important.

To date, there are/have been multiple international anti-terrorism wars led by the United States, which seem to be the initial step for establishing the global anti-terrorism regime. The war against terrorists is an actual military conflict that aims for the complete destruction of terrorist groups and their regimes. Terrorists, of course, fight back with *asymmetric warfare*. For political purposes, terrorist groups take irregular challenges, occasionally attacking civilians, with the aim to instill panic or fear. Since many terrorist attacks are rather unexpected from our side, it is very difficult to predict exactly when or where we could be attacked, and the regular deterrent strategy against the obvious opponent is useless.

The United States therefore uses the *Preemptive Strike* strategy. The George W. Bush Administration asserted both a right and the intention to wage preemptive or preventive wars. This became the basis for the Bush Doctrine which ultimately weakened the unprecedented levels of international and domestic support for the United States, which the September 11 attacks had garnered. Beginning with his State of the Union address on January 29, 2002, Bush began to publicly focus attention on Iraq, which he labeled as a part of an *axis of evil* allied with terrorists, and on how Iraq posed "a grave and growing danger" to US interests through its possession of WMDs. In the latter half of 2002, CIA reports contained assertions of Saddam Hussein's intent of reconstituting nuclear weapons programs and not properly accounting for Iraqi biological and chemical weapons, and that some Iraqi missiles had a greater range than allowed by UN sanctions. Contentions that the Bush Administration manipulated or exaggerated the threat and evidence of Iraq's WMD capabilities would eventually become a major point of criticism for the President.

Another problem related to the "preventive action" approach, is its interference with a nation's sovereignty. Then there is the issue regarding rendition aircraft. *Rendition aircraft* are aircraft used by national governments to move prisoners internationally. Apart from investigations concerning the extraordinary rendition program (European Parliament February 2007 report), which concluded that the US CIA had operated 1,245 flights on European territory, several European countries have opened specific investigations concerning CIA flights. For example, on December 2, 2005, conservative newspaper *Le Figaro* revealed the existence of two CIA planes, suspected to have been transporting CIA terrorist prisoners back to the United States, that had landed in France. Though these aircraft operations were related with "anti-terrorism action", these irregular operations are problematic for the European nations.

Future strategies against terrorism will be a suitable combination of (1) a strong anti-terrorism regime and (2) an anti-terrorism war. Once a terroristic action occurs, as a penalty for violating the anti-terrorism regime, strong military force should be used in response.

7.5.2 *Natural Disaster*

Natural disasters include earthquakes, volcanic eruptions, tsunamis, tornadoes, hurricanes, floods, lightning, etc. Figure 7.7 shows the amount of natural disaster damages sustained by various areas around the world (in ratio to the total amount of US$ 1.1 trillion, accumulated from 1970 to 2004) [11]. The United States is shown to suffer from the most damages caused by natural disasters, followed by Japan. These statistics are primarily proportional to the respective country's GDP amount. It also shows that the Asian area (including Japan) unfortunately bears nearly half the costs of the world's disaster damages due to earthquakes and typhoons.

In the US Federal Emergency Management Agency (FEMA), the Emergency Management and Response — Information Sharing and Analysis Center (EMR-ISAC) promotes *critical infrastructure protection (CIP)* by sharing their CIP details and emerging threat information with Emergency Services Sector (ESS) departments and agencies nationwide. The ESS includes emergency management, fire and emergency medical services, hazardous materials teams, law enforcement, bomb squads, tactical operations/special weapons assault teams, and search and rescue.

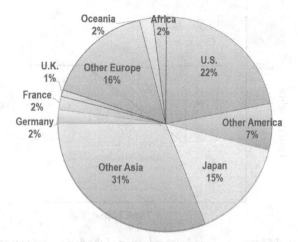

Fig. 7.7.   The amount of natural disaster damages sustained by various areas (in ratio to the total amount of US$ 1.1 trillion, accumulated from 1970 to 2004).

Basically, we need to develop a "disaster resilient society." The Multidisciplinary Center for Earthquake Engineering Research (MCEER) at the State University of New York at Buffalo, USA, defined *disaster resilience* as the "reduced probability of system failure, reduced consequences due to failure, and reduced time to system restoration." These three desired outcomes constitute the essence of the framework proposed by MCEER to quantitatively define resilience. Figure 7.8 visualizes the concept of "Resilience," which can be evaluated by the triangular area in the figure. The functionality of the infrastructure (vertical axis) is plotted as a function of time lapse (horizontal axis). Once the disaster occurs, the functionality diminishes suddenly; however, this sudden reduction should recover as a function of time. Minimizing the triangular area of this crisis dip is the key to enhancing resilience. Compared with a normal recovery situation, Fig. 7.8 inserted two different situations: (a) reduction of the damage amount and (b) reduction of the recovery time. The former corresponds to the disaster mitigation model, while the latter corresponds to the engineering dominant model.

There seems to be *four-Rs* for enhancing resilience: *Robustness, Redundancy, Resourcefulness,* and *Rapidity*. Buildings should be structurally strong and robust enough to withstand an earthquake. Infrastructures should have multiple redundancies so that the reduction in functionality can be maintained by backup systems in the event of a disruption. In order to accelerate the speed of recovery, decision-makers must give adequate consideration to how the necessary manpower, rescue items, money, disaster damage information, and crisis technologies can be collected. Finally, rapidity may be obtained by a suitable *crisis management organization*, in addition to the above resources.

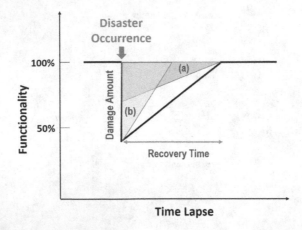

Fig. 7.8.   Concept of "disaster resilience": (a) reduction of the damage amount and (b) reduction of the recovery time.

7.5.3 *Nuclear Power Plant Accident*

There have been three nuclear meltdowns due to nuclear power plant accidents in the 50 years since the first commercial application of atomic energy. The affected plants were: Three Mile Island (USA), Chernobyl (Russia), and Fukushima Daiichi (Japan). While I do not claim to be an expert in nuclear engineering; however, I have had practical experience in three of the world's most famous nuclear incidents. The only one I did not personally experience was the Chernobyl incident. I grew up in Hiroshima in the early 1950s, in the area previously damaged by the explosion and after-effects of the Little Boy atomic bomb in 1945. Then, as a young assistant professor at Tokyo Tech, Japan, and after joining Penn State University, USA, during my sabbatical, I assisted in the development of new cement to shield the nuclear reactors damaged in the Three Mile Island incident in 1979. I was also involved in the Fukushima Daiichi incident; as a Navy Ambassador to Japan setting up the Japan–US Agreements in the rescue programs during my sabbatical from Penn State more than 30 years later. These experiences have led me to feel many emotions about issues pertaining to nuclear power plant security and safety.

7.5.3.1 *Three Mile Island vs. Fukushima Daiichi Nuclear Power Plant Accidents*

Table 7.6 shows the status comparison between the Three Mile Island and Fukushima Daiichi nuclear power plant accidents. The major difference is the explosion, which is discussed below.

Table 7.6. Comparison between the Three Mile Island and Fukushima Daiichi nuclear power plant accidents.

	Three Mile Island (USA)	Fukushima Daiichi (Japan)
Date	1979.3.28	2011.3.11
Reason	Human error of equipment usage	Earthquake/Tsunami and Electrical shutdown
Level	Melt-down	Melt-down and Explosion
Command	President/State Army	TEPCO/Prime Minister
Initial Actions	Water spray/ Evacuation in 8 km area in 24 h	Delay of water spray/ Evacuation in 10 km area in 16 h

The Three Mile Island accident was a partial nuclear meltdown that occurred on March 28, 1979, in one of the two Three Mile Island nuclear reactors in Dauphin County, Pennsylvania, United States. It was the worst accident in US commercial nuclear power plant history [12]. This incident was rated a "5" on the seven-point International Nuclear Event Scale; i.e., an "Accident with Wider Consequences."

The accident began with failures in the non-nuclear secondary system, followed by a stuck-open pilot-operated relief valve in the primary system, which allowed large amounts of nuclear reactor coolant to escape. The mechanical failures were compounded by the initial failure of plant operators to recognize that the situation had been caused by the accidental loss of coolant due to inadequate training. Other human factors, such as human–computer interaction design oversights and ambiguous control room indicators in the power plant's user interface also played a part. In particular, a hidden indicator light led an operator to manually override the automatic emergency cooling system of the reactor because the operator had misunderstood the activation of the emergency cooling system as an indication that there was too much coolant water present in the reactor and that they were causing the release of steam pressure. The reader should note that electronic sensors and actuators, as well as the computer technologies needed in good nuclear plant safety systems, had yet to be fully developed in 1979. By the evening of March 28, the core appeared to be adequately cooled and the reactor appeared to be stable. But new concerns arose by the morning of 30th March, Friday. A significant release of radiation from the plant's auxiliary building to relieve pressure on the primary system and avoid curtailing the flow of coolant to the core, caused a great deal of confusion and consternation. The partial meltdown resulted in the release of unknown amounts of radioactive gases and radioactive iodine into the environment. In an atmosphere of growing uncertainty about the condition of the plant, the Governor of Pennsylvania, *Richard L. Thornburgh*, consulted with the NRC on evacuating the population near the plant. Eventually, he and NRC Chairman *Joseph Hendrie* agreed that it would be prudent to evacuate members of society near the plant most vulnerable to radiation. Thornburgh made a public announcement advising pregnant women and children below five years of age residing within an 8 km radius of the plant to leave the area.

Within a short time, chemical reactions in the melting fuel created a large hydrogen bubble in the dome of the pressure vessel, which is the container that holds the reactor core. NRC officials worried that the hydrogen bubble might burn or even explode and rupture the pressure vessel. In that event, the core would fall into the containment building and perhaps cause a breach of containment. The hydrogen bubble was a source of intense scrutiny and great anxiety, both among government authorities and the local residents, throughout the day on Saturday, March 31. The crisis ended when experts determined on Sunday, April 1, that the bubble could not burn or explode due to the absence of oxygen in the pressure vessel. Furthermore, by that time, the utility had succeeded in greatly reducing the size of the bubble.

The Fukushima Daiichi nuclear disaster occurred after a 9.0 magnitude earthquake and subsequent tsunami in the Tohoku region, Japan, on March 11, 2011. The chronicle of this disaster is provided below [13]:

Friday, March 11

- 14:46: The earthquake struck off the northeast coast of Honshu Island at a depth of about 24 km from sea level. The Fukushima Daiichi (I) power plant's nuclear reactors 1, 2, and 3 were automatically shut down due to the tremor. Nuclear reactors 4, 5, and 6 were undergoing routine maintenance and were not in operation. The tremor had the additional effect of causing the interruption of supply from the Japanese electricity grid to the power plant; however, backup diesel generators kicked in to continue cooling. Tokyo Electric Power Company (TEPCO), the plant's operator, found that units 1 and 2 were not operating properly and notified the respective officials.
- 15:14: The Nuclear and Industrial Safety Agency of Japan (NISA) initiated emergency headquarters in an attempt to gather information on the 55 nuclear reactors in Japan.
- 21:23: An evacuation order was issued by the government to people living within a radius of 3 km from the Fukushima I station. Those within a radius of 10 km were told that they could remain in their homes. TEPCO announced that the pressure inside the reactor for unit 1 of Fukushima I was more than twice its normal level.

Saturday, March 12

- 1:30: TEPCO requested for permission to open the vent of Fukushima I, and the Government approved.
- 6:14–8:04: *Prime Minister Naoto Kan visited* Fukushima Nuclear Power Plant.
- 8:27: Evacuation from Okuma town was confirmed.
- 9:04: The vent was tried.
- 14:53: 80 ton of water was injected to cool the reactor of Unit 1 of Fukushima I.
- 15:36: A massive *explosion* occurred in the outer structure of unit 1. Evacuation of people residing within 10 km of Fukushima I and of people residing within 3 km of Fukushima II was declared.
- 19:04: Seawater injection into reactor 1 was started. TEPCO ordered Daiichi to cease seawater injection at 19:25, but the head of the Daiichi plant, Masao Yoshida, ordered workers to continue with the seawater injection.
- 20:32: The evacuation zone around Fukushima I was extended to 20 km, while the evacuation zone around Fukushima II was extended to 10 km.

Monday, March 14

- 11:01: The reactor building of unit 3 exploded. According to TEPCO reports, there was no release of radioactive material beyond the point that was already being vented, but blast damage had affected the water supply to unit 2.

Tuesday, March 15

- 6:00: The reactor of unit 2 exploded, which damaged the area on the 4th floor above the reactor and spent fuel pool of the reactor of unit 4.

Thursday, March 17

- 9:48: During the morning, Self-Defense Force helicopters dropped water four times on the spent fuel pools of units 3 and 4.

7.5.3.2 *Key Reasons to Nuclear Power Plant Explosion*

Commanding System

Once the Three Mile Island (TMI) incident happened in the United States, the commanding role was taken by the Governor of Pennsylvania, *Dick Thornburgh* (commander-in-chief) and the State Army, with *James Floyd* (supervisor of TMI-2 operations) as advisor. President *Jimmy Carter* toured the TMI-2 control room on April 1, 1979, with NRR Director *Harold Denton* to provide advice to the Governor on major issues, including the emergency budget amount. The evacuation of the surrounding people was then smoothly conducted by the State Army.

In contrast, in the Fukushima Daiichi (II) Nuclear Plant disaster, the commanding system was not clear due to three major reasons: (a) Prime Minister *Naoto Kan* tried to convince the TEPCO chief executive to take the position of commander-in-chief. But, a president of a single company may not have complete details about the crisis incident. There were numerous incidents of miscommunication between the TEPCO executive and subordinates, as exemplified by this log entry: "At 19:04, seawater injection into reactor 1 was started. TEPCO executive ordered Daiichi Plant director to cease seawater injection at 19:25, but Director *Masao Yoshida* ordered workers to continue with the seawater injection; neglecting his boss' order." Though Mr. Yoshida's decision later proved to be the better choice, ignoring the decision of his superior would have been considered an act against the system in the military. (b) Prime Minister *Naoto Kan's* visit to the Fukushima Nuclear Power Plant that morning caused a serious *delay in the "pressure vent"* decision and triggered the hydrogen explosion of unit 1. (c) The evacuation of people residing within the 3 km radius from the plant took such a long time; resulting in the major delay for the vent action. This is mainly because military forces were not used to facilitate the evacuation.

Lessons: (a) A company like TEPCO primarily prioritizes earning a profit for providing electrical power at a reasonable price to the public, over public welfare. Hence, it is difficult to force them to consider a crisis security issue like this power plant accident that resulted from a natural disaster. The Japanese national/local government should take the commanding responsibility immediately after the crisis happens. (b) The commander-in-chief should be the one keeping the commanding address consistent (no confusion in the order). The commanding officer needs to be able to decide immediately during the crisis situation, even without precise/enough information. (c) They should use the military/self-defense force for the evacuation process; not voluntarily, but mandatorily.

Technological Problem

After the Three Mile Island (TMI) accident, the power plants in the regions of US and Europe (except Japan) updated and replaced their (1) water height sensors (which was a bottleneck), (2) backup manual devices for "vent valve control" (longer shaft for rotating the valve from a long distance), and (3) sophisticated filters at the vent part, which traps most of the nuclear waste. Unfortunately, the

Japanese had not installed any of these three tools, which was considered the main reason for the higher level of contamination experienced in the Fukushima incident. The Japanese people were generally uninterested in developing new military or crisis technologies after World War II. During my tenure as a Japanese university professor until the 1980s, the development of rescue robots had been popular, and most of the nuclear power plants did have a rescue robot for crisis situations. However, in the last quarter of a century, the Japanese government eliminated the use of rescue robots intentionally, so as to shape the perspective of the general public toward nuclear technology; to remove the idea of that a crisis could happen and/or to promote the belief that nuclear technology is perfectly safe. Japan did not even have "Drones" or "Unmanned Underwater Vehicles," which are modern day basic necessities in search and rescue operations; though their robotic technologies are rather advanced in industrial and entertainment applications, as exemplified by the Honda robot "ASIMO," a compact and cute toy, unusable for rescue purposes.

I was personally involved in a new cement development project a couple of weeks after the TMI incident which took place on March 28, 1979. The then-director of Materials Research Laboratory (MRL), the late *Rustum Roy*, immediately wrote a proposal and received a US$1 Million (equivalent to US$10 M now) research fund from the Federal Government. Since the land in Pennsylvania includes plenty of rock-salt, we needed to develop a special cement which solidifies quickly and stably to shield the melted down nuclear reactor completely. Though I was not supposed to work on the cement development (I had originally been engaged to work on piezoelectric ceramics), 25% of my research time was spent on this new urgent project and I helped with the analysis of X-ray crystallography intensively for three months. A three-to-six-month period seems to be a reasonable duration within which to develop this sort of crisis technology in the US.

In contrast, the development of crisis technology in Japan was very slow even after the Fukushima incident. No reliable rescue robot was developed in Japan even after half a year had passed — such that the US military rescue robots (made by iRobot) had to be used for the monitoring and inspection of the nuclear plant's structure. The Japanese government did not conscript·eminent robotic engineers as a mandatory duty even for this emergency crisis occasion.

Lessons: The Japanese government needs to reinvest in developing backup, rescue systems for all nuclear power plants, and also create a risk/crisis management system to conscript the necessary engineers to develop the urgent crisis technologies, by setting up an emergency research fund. The commander-in-chief should be an expert on collaborating effectively with military/self-defense forces, and not just a company executive. Crisis management is out of a private company's power.

Evaluation by the Public

Twenty-eight hours after the TMI accident began, *William Scranton III*, the lieu- tenant governor, appeared at a news report briefing to say that *Metropolitan Edison*, the plant's owner, had assured the state that "everything is under control." Later that day, Scranton changed his statement, saying that the situation was "more complex than the company first led us to believe." There were conflicting statements about the release of radioactivity. Schools were closed and people were urged to stay indoors. Farmers were told to keep their animals under cover and on stored feed. Governor *Dick Thornburgh*, on the advice of NRC chairman *Joseph Hendrie*, advised the evacuation "of pregnant women and pre-school age children ... within an 8 km radius of the Three Mile Island facility." The evacuation zone was extended to a 32 km radius on Friday, March 30. Within days, 140,000 people had left the area. More than half of the 663,500 population within the 32 km radius remained in that area. According to a survey conducted in April 1979, 98% of the evacuees had returned to their homes within a period of three weeks.

Post-TMI surveys have shown that less than 50% of the American public were satisfied with the way the accident was handled by Pennsylvania State officials and the NRC, and people surveyed were even less pleased with the utility provider (General Public Utilities) and the plant designer.

In contrast, the Japanese Democratic Party government was elusive in providing complete information and security to the people during the Fukushima accident, thereby losing the trust of the Japanese people towards the incumbent government. This resulted in the Democratic Party's loss in the next election held in December, 2012. Some of the bad fashions of hiding the necessary information from the public include:

- On the morning of March 12 (one day after the Fukushima Daiichi accident), *Koichiro Nakamura*, Counselor of Nuclear and Industrial Safety Agency (NISA) mentioned that "the meltdown might have happened in the reactor" during an interview with a reporter. However, the government soon after expelled this person from the spokesman's role, and never mentioned the possibility of a meltdown again until the explosion happened.
- At 8:00 am of March 12, NHK (Japan Broadcasting Company) TV news released the phone conversation between a US DOD executive and a Japanese Government executive, in which a voice shouted, "Spray water immediately, since cooling is most important." Though this kind of important news is usually repeated in the noon program, NHK never broadcasted the recorded phone conversation again. The government might have used its political

power to stop the news, because the government was not thinking about spraying water to cool the reactor at that time.

- Secretary of Nation, *Yukio Edano*, did not use the term "hydrogen explosion" during the first news release. The term "explosion-like incident" was used instead, in order to camouflage the severe incident, until foreign news channels broadcast the apparent explosion videos. They reported that "Only the Japanese government hesitates to admit this serious accident."
- The Nuclear Safety Division of the Ministry of Education, Culture, Sports, Science and Technology (MEXT) streams information from a national network of detectors, called the System for Prediction of Environment Emergency Dose Information (SPEEDI). It has been called a "computer-based decision support system" by researchers, and its function is for real-time dose assessment in radiological emergencies. Actual aerial monitored data on the dispersal of radioactive materials were collected by US forces every day and provided to the Japanese MEXT immediately after March 11. The Japanese public only got these official data almost two weeks later, on March 23, 2011. Calculations were made with SPEEDI by the science ministry and NISA by assuming the amount of radioactive substances, and these results could have helped local governments and people in their choice of more appropriate evacuation routes.
- In reality, the evacuees initially moved in the same direction in which the radioactive materials dispersed owing to the wind and geological conditions, merely because the government did not disclose the above SPEEDI predictions. They needed to relocate again a couple of days later, after detecting a high "dose" rate in that area. Because of unreliable and untrustworthy crisis management by the Japanese Democratic Party government, US President Obama ordered the mandatory evacuation of the family members of US Navy Officers from Japan within 72 h on March 17.
- Because of this mishandling of the Fukushima Daiichi Nuclear Power Plant accident, the Democratic Party lost the election held in December 2012.

Lessons: The Japanese government needs to avoid hiding necessary information from the public in such crisis situations, and needs to act quickly by first resolving the most urgent issue at the time. In the case of the Fukushima disaster, they should have "spray[ed] water immediately!"

7.5.3.3 *Nuclear Power Plant Future Situation*

Figure 7.9 shows the trend of world nuclear reactors in operation over the years, in terms of number of units and updated capacities in GW [14].

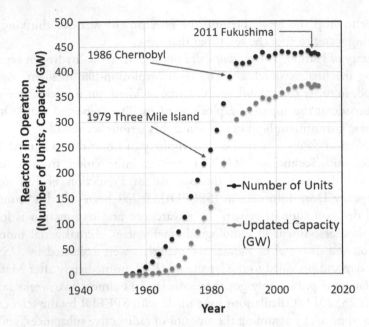

Fig. 7.9.　Trend of world nuclear reactors in operation over the years (number of units and updated capacities in GW).

The United States ranks first with 104 reactors, followed by France with 64 reactors (which cover 75% of the nation's electric energy). The third is Japan, which has 50 reactors (54 reactors before the Fukushima incident). From Fig. 7.9, the reader can identify a drastic deceleration in dependence on nuclear reactors after the Chernobyl accident, and an actual reduction in the use of reactor units after the Fukushima incident. However, a complete termination of the use of nuclear power plants seems to be unrealistic owing to two key issues: (1) global warming/greenhouse gas reduction, and (2) the price of electricity charged to the public.

The "Energy-Mix Plan 2015" alternative proposed by the Japanese government is introduced in Fig. 7.10 [15]. The government committee report (chaired by *Masahiro Sakane*, Komatsu) described that:

(1) Replacing the nuclear power plants with renewable energy methods such as solar cells, windmills or hydroelectric power systems is not realistic in Japan from the viewpoint of the electrical cost it would incur. Because solar cells do not effectively or constantly generate power due to the sun shine rate, its maximum contribution to the Japanese power grid would be limited to

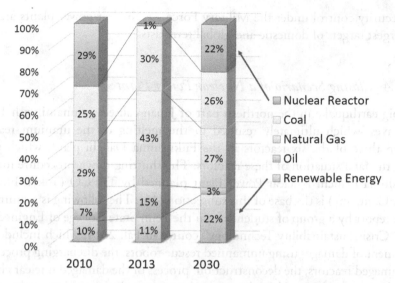

Fig. 7.10. "Energy-Mix Plan 2030" proposed by Japanese government (May 27, 2015).

7.0%. Windmills are riskier because their exposure to many typhoons is likely to damage the system. They could also cause a great disturbance to maritime or fishing activities. An estimated 1.7% contribution by windmills to Japan's energy needs seems to be a reasonable estimate.

(2) Owing to the hurdle of the global CO_2 reduction regime, fossil fuel (*oil, natural gas, coal*) usage should be limited to 3%, 17%, and 26%, respectively. Accordingly, the remaining 22% of electricity should be generated by the nuclear reactors. Compared to the situation in the pre-Fukushima Accident period, this reflects a 25% reduced dependence on nuclear power (from 29% to 22%).

Suggestions: I think that the Japanese government's "Energy Mix Plan 2015" scenario seems to be reasonable. However, in order to re-operate the current existing 50 reactors, improvements to their security and safety systems must be made; such as:

(1) Improvements in their monitoring sensor system — Redundancy is important.
(2) Rescue robot system installation — During sudden (or) unexpected accidents due to natural disasters, unmanned autonomous robots are mandatory.

(3) Security control under the Military Force — Nuclear power plants are the largest targets of domestic and global terrorists.

7.5.3.4 *Closing Scenario of a Nuclear Power Plant*

The big earthquake in the northern part of Japan caused a tsunami with 10 m tall waves, which ultimately resulted in the melting of the uranium reactor rods in three of the six reactors in the Fukushima Daiichi plant, which then led to the fatal situation of these reactors. The shutting-down procedure for the Fukushima Daiichi Nuclear Power Plant planned by TEPCO (Tokyo Electric Power Company) is the base of this subsection [16]. The following is a summary of the report by a group of students from the Penn State College of Engineering for a "Crisis/Sustainability Technology" course in Fall 2014, which includes an assessment of damages using unmanned rescue robots, the discharging process of the damaged reactors, the deconstruction process of the damaged nuclear chambers, the intermediate and final storage place for nuclear waste, research on the methods of reactor storage, research to accelerate nuclear reactions, contact with related governmental offices and shut down cost estimates. I would particularly like to acknowledge the efforts of one of the students, *Steven Aloia*.

Inspection

Before work can begin on the reactor sites, unmanned robots must survey each reactor, as the area is too *radioactive* for human occupancy. In some areas, the dose rate is 54 mSv per hour, which is a year's allowable dose for a cleanup worker. The first to examine the reactor areas will be the inspection robots. The robots will be equipped with both cameras and dosimeters to identify radioactive hot- spots. Due to the nuclear plant's massive steel and concrete structures, the use of wireless communication will be ineffective. The robots must therefore be tethered by a cable. The robots must not only be able to unspool the cable behind them but must also be able to automatically take up the slack when they change direction, to guard against tangles. The robots must be waterproof enough to roll through puddles, and must be able to carry heavy cameras capable of detecting *gamma radiation*. When the robot is exposed to the radiation, it too will become radioactive; therefore, it must be able to recharge itself by connecting to a power source [17].

After these robots finish the initial inspection of the nuclear chambers, the data will be used to create a map of the debris and nuclear hotspots within each of the reactors. The second stage of the decommissioning process will be to use the data acquired by the inspection robots to decontaminate the nuclear

Fig. 7.11.   Hitachi ASTACO-SoRa demolition robot used to clear debris [18]. [Courtesy of Hitachi]

hot-spots. This decontamination process will require two other classes of robots. In order to clear a path for the decontamination robots, demolition robots equipped with interchangeable tools in each arm such as grippers, cutting blades, and drills will be used to remove the debris (Fig. 7.11) [18]. After the areas are cleared, yet another class of robots will be used to decontaminate the floors and walls using high-pressure water and dry ice to abrade the surfaces. These robots will scour away radioactive materials along with top layers of paint or concrete and vacuum up the resulting sludge.

Discharging Process of the Reactors

Within each reactor building is a holding pool, which is designed to hold the spent fuel rods until they are moved off site. The next step of the decommissioning process is to remove these spent fuel rods from the still-intact pools and transport them to another on-site holding pool. The process will be repeated for each of the damaged reactors (1, 2, 3, and 4) and undamaged reactors 5 and 6 (less challenging). Before the removal of the spent fuel rods, the damaged reactor buildings will be retrofitted with an exoskeleton which will support the overhead crane used to remove the spent fuel assemblies. After the retrofit, the removal of spent fuel assemblies will proceed as outlined in Fig. 7.12 (based on TEPCO's roadmap [16]).

(1) Fuel Assembly Containment — Grab fuel assemblies stored in fuel rack and
     load them into cask in the cask pit underwater.

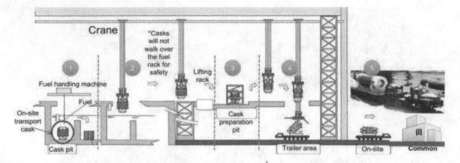

Fig. 7.12.   Fuel assembly removal process, by using the systems with the same structures, design and safety as regular fuel handling. The cask pit transport pictured has a good track record, having successfully completed more than 1,200 rounds of tests at TEPCO.

(2) Cask Preparation — Lift the cask from the cask pit and transport it to the cask preparation pit on the lifting rack inside the cover.

(3) Decontamination — Seal lid in the cask preparation pit and perform decontamination of cask.

(4) Load — Lift the cask by crane, put it down in the trailer area and load it onto the transport vehicle.

(5) Transport — Transport the cask to the on-site common pool.

Radioactive Water Containment

The next step of the closing process is to contain the radioactive water that flows freely through the site. Approximately 400 tons of groundwater per day streams into the basements of Fukushima Daiichi's reactors, it then mixes with radioactive cooling water from the leaking reactor vessels. In the interim, on-site collection tanks will be constructed to hold the radioactive water until the leaks are fixed. In order to stop the flow of groundwater, a series of underground pipes will be constructed. The pipes will then be used to transport fluid through in order to freeze the ground around the leaking reactors. The *frozen ground* will act as a wall to contain the radioactive water.

Stored Water Purification and Release

After containing radioactive groundwater seepage via the frozen wall containment, *purification filter systems* will be installed on-site to bring the stored radioactive water to a safe level with respect to radioactive materials. The resulting purified water will be limited to trace amounts of tritium which is a radioactive isotope of hydrogen. Tritium is less dangerous than other radioactive materials because it passes quickly through the body. After the levels have been tested to meet the passing amount of tritium, the water will then be released into the Pacific Ocean where it would have a negligible effect.

Reactor Leak Location Detection and Fix

The leaks in the reactor vessels are the main modes of failure as well as what caused the contamination of groundwater. After the groundwater has been contained, the next critical step is to find the location of the leaks within the reactor vessel and address them.

Many of the reactor leaks are thought to be in the suppression chambers. The suppression chambers are doughnut-shaped structures that cover the containment vessel and typically hold water, thereby regulating the temperature and pressure inside the pressure vessel during normal operations. Before these leaks can be addressed, they must first be located. Due to the fact that these suppression chambers are currently filled with water, the process of leak detection is a large challenge. The development of specialized robots will be necessary to complete this task. The robots will need to be able to move through the narrow pipes of the containment vessels by changing shape, while another version of this robot will take on the challenge of moving through radioactive water to perform research related to the task of removing melted nuclear fuel. The robots need to be able to move around inside the containment vessels of the reactors that experienced meltdowns, and whose dismantling poses never-seen-before engineering challenges.

Once the robots have detected the leaks, work can begin on sealing the leak points. The exact type of leakage mending technology needed will depend on the type and location of the leak. One possible fix would be to use a *machine hatch*, which can be put in place and then tightened against the walls of the dry well like a jack to seal the fixture. A second option would be to use an *air lock* to, again, seal the wall and tighten the fixture using a turnbuckle. Both of these options can be remotely installed through the use of robots. If this process of detecting leaks and repair proves to be impractical, a second, more crude and simplified option is available. It is possible to pour concrete into the suppression chamber and the pipes that lead to it. By taking this approach, it is possible to make a seal between the containment vessel and the suppression chamber, at which points the leaks will prove not to be critical. Regardless of the chosen method, it is necessary to seal the leaks as a precautionary measure before the removal of the damaged nuclear cores.

Removal of Damaged Cores

The first step of core removal would be to flood the containment vessel to the top so that the water will act as a shield against the radioactive fuel. After the damaged core is fully submerged, submersible robots will map the slumped fuel assemblies within the pressure vessels. The robots used for this task will be created by adapting those used by the petroleum industry to inspect deep-sea oil wells. After completing this process, drills will be used to reach the bottoms of the cores (25 m down). Once the drills reach the bottom of the pressure vessel, they will then break up the metal pooled there from the melted reactor rods. The pooled metal (corium) is expected to have a hard crust with softer material inside. In order to design the proper tooling, tests will be conducted using computer models as well as lab tests using corium. Once broken up, the core material will be contained in casks and removed.

Intermediate and Final Storage Location

As radioactive debris is removed from the Fukushima site, it is necessary to find both an intermediate (30 years) and final storage location. The intermediate storage location has been proposed to be built between the two towns of Futaba and Okuma. The proposed area for the facilities will cover approximately 16 sq. km. It is understood that more than 2,000 people own pieces of land in the proposed storage area. Officials will have to negotiate compensation deals with each of them. It was proposed to offer to lease the land so that the former residents will not have to lose their property because, unlike American culture, Japanese prefer to stay in their own homes/on their own lands, which they see as a heritage from their ancestors. This usually causes the negotiation period to stretch — sometimes longer than several years.

Looking forward the next 30 years, a final storage location for Fukushima's nuclear waste will still need to be determined. Due to Japan's limited space and high seismic activity, this task will not be easy. Studies are currently underway in Horonobe, Hokkaido, which is located near Japan's northernmost point. As time progresses, new technical advancements may lead to a better permanent storage answer.

Research to Reduce Nuclear Half-Life

A critical step in dealing with the waste left behind by Fukushima along with all nuclear waste is the time that these wastes take to become non-radioactive. Reducing this time will allow for nuclear waste to be disposed of much more rapidly, thus reducing the need for permanent storage areas. One promising means of accomplishing this task is by supercooling radioactive waste. Experiments conducted by *Claus Rolfs* of Ruhr University in Bochum, Germany, have proven that this process is feasible [19]. Rolfs fired radioactive nuclei, encased in metal and cooled, into a particle collider to see whether the free electrons could accelerate the ejection of positively charged particles from a radioactive nucleus. As expected, he observed that radioactive decay occurred and was accelerated considerably by the presence of the lower temperatures and metal casing. According to Rolfs, the technique could potentially reduce the half-lives of radioactive material by a factor of 100 or more. With continued research, it is possible that this method could be the solution for the age-old question of how to dispose of nuclear waste.

Government Relations

The Japanese government should monitor all steps of the Fukushima closing process. The most critical government relations will come during the stages of radioactive wastewater purification standards. Once the radioactive wastewater levels have reached a low enough level, the water will be released into the Pacific Ocean. Although this will have a negligible effect on the region's ecology, it is a critical public relations topic for which the backing of both the local and national government will be necessary. The other critical government approval comes in the form of the storage of the nuclear waste. It will be necessary to work with both local and national governments for them to establish and approve both interim and final waste storage and disposal locations.

Cost Estimate

As with any unique project, the cleanup of Fukushima requires a vast amount of money to be invested into the R&D work to develop solutions to never- before-seen problems. The total cost of the clean-up and closing of Fukushima is estimated to be between US$150 billion and US$250 billion. The majority of the funds will come from the Japanese government. As the project continues over the next 40 years, this cost estimate is subject to change based on unforeseen problems that will need to be overcome.

Education of Engineers

During the past 20 years, the Japanese government and Japanese universities discouraged academics from pursuing research in, and from teaching courses on, nuclear engineering and rescue robots; thus, leading to the dramatic reduction of the number of experts and students in this area — particularly after the Fukushima Accident. It is essential for Japanese society to intentionally cultivate nuclear engineers in order to develop the above-mentioned technologies in this coming half a century. Increasing the number of expert professors/labs in the Nuclear Engineering Department in universities by considering special research funds from Japan's Ministry of Education, Culture, Sports, Science and Technology (MEXT) seems to be important.

7.6 Summary

In this chapter, crisis management methods were discussed in terms of the levels of threat provider and receiver. These levels are: (1) nation/state, (2) non-nation group (e.g., domestic rebels, international terrorists, etc.), and (3) non-human (e.g., natural disasters, infectious diseases, etc.). "Decide → Conduct → Observe → Validate" was presented as a cycle of action decision in crisis management, in general. I also talked about the "four Rs" for enhancing the resilience in institutes, societies, or states against various threats. The four Rs are: *Robustness, Redundancy, Resourcefulness,* and *Rapidity.*

The lessons learned from the Fukushima Daiichi Nuclear Power Plant accident are important. I established that a company like TEPCO prioritizes company profit for providing electricity service at a reasonable price to the public over public welfare, and how that consequently makes it difficult to force them to consider crisis security issues like this power plant accident. I argued that the Japanese national/local governments should take the commanding responsibility immediately when a crisis like the Fukushima nuclear disaster happens. I believe that the *commander-in-chief should be the one* keeping the commanding address consistent; that he/she needs to make decisions immediately during the crisis situation — even without precise/enough information; and that he/she should use the military/self-defense forces in order to achieve a more effective/efficient evacuation process.

I showed how the Japanese government's "Energy Mix Plan 2015"s decision to *keep Japan's nuclear reactors generating 22% of the nation's electricity by 2030,* is reasonable. However, I also presented how, in order to re-operate the currently existing 50 reactors, improvements to existing security and safety systems are needed. The proposed improvements included improvements in monitoring sensor systems, the installation of rescue robot systems, and how control over the plants' security should be given to the nation's Military Force because nuclear power plants are targets of domestic and global terrorists. I also elaborated on why a closing scenario of a nuclear power plant should be deeply discussed with the society it affects.

7.7 Problem

Choose one topic most related with the reader's local area from the following crisis list, and discuss the possible scenarios the way we discussed the "Closing Scenario of a Nuclear Power Plant" in Section 7.5.3.4.

Crisis List: (a) Tornado
              (b) Hurricane/Typhoon/Cyclone
              (c) Volcanic eruption
              (d) Medical epidemic
              (e) Terrorist attack on the power plant

Scenario: (1) Prediction/forecasting, surveillance/monitoring
              (2) First action items — Information Transfer, people's evacuation, etc.
              (3) Second action items — Protection methods, protection technologies (shelters, firearms), immunization, etc.
              (4) Situation monitoring
              (5) Closing decision/remark — Public announcement
              (6) Future homework learned from the experiences

References

[1] Yoshinobu Yamamoto, *Security Mechanism on Asian Pacific*, Sairyu Pub (2005).
[2] http://www.wsj.com/articles/SB120571689020540529.
[3] https://en.wikipedia.org/wiki/Arab_Spring.
[4] K. Uchino, *Entrepreneurship for Engineers*, CRC Press, New York, NY (2010).
[5] https://en.wikipedia.org/wiki/In_Amenas_hostage_crisis.
[6] http://en.wikipedia.org/wiki/Sierra_Leone_Civil_War.
[7] Department of the Army, "FM 3-0 Operations Manual", Feb 2008. http://usacac.army.mil/cac2/Repository/Materials/FM3-0(FEB%202008)/pdf.
[8] http://www.econlib.org/library/Knight/knRUPCover.html.
[9] http://en.wikipedia.org/wiki/Al-Qaeda.
[10] http://www.dhs.gov/publication/advise-report.
[11] http://www.cred.be.
[12] http://www.nrc.gov/reading-rm/doc-collections/fact-sheets/3mile-isle. html#summary.
[13] https://en.wikipedia.org/wiki/Timeline_of_the_Fukushima_Daiichi_nuclear_disaster.
[14] Data taken from IAEA: Power Reactor Information System, 2014. https://www. iaea.org/pris/.
[15] Asahi Newspaper, May 27, 2015, p. 5.
[16] http://www.tepco.co.jp/en/nu/fukushima-np/roadmap/images/d141030_01-e.pdf.
[17] http://www.iros2012.org/site/sites/default/files/iros2012_HajimeAsama_Plenary.pdf.
[18] http://fukushimaupdate.com/hitachi-develops-robots-to-probe-fukushima-no-1-plant/.
[19] http://aussiethule.blogspot.com/2006/10/half-life-heresy-accelerating_23.html.

CHAPTER 8

Advanced Game Theory

Historically, crisis technologies in advanced countries were developed domestically and confidentially. However, because the characteristics of crises have changed significantly — for example, the threat by global terrorist organizations such as al-Qaeda (AQ) and the Islamic State of Iraq and Syria (ISIS) — global/international collaboration with respect to both political and Science and Technology (S&T) aspects is highly required. In this chapter, we consider how to effectively establish international relationships for crisis and sustainability developments, particularly from the S&T viewpoint.

8.1 Four Types of International Security Systems

Global security systems are considered in this section from the viewpoint of threat provider and receiver. Table 8.1 shows four types of global security systems [1] discussed from a game theory viewpoint. Review the game theory fundamentals in Section 3.3 prior to reading forward.

8.1.1 *Competitive Security System (Type A)*

The threat group could be identified as one that exists externally (Cell A in Table 8.1), so that the internal group members are bound tightly and create a regime. This situation occurs when the global systems are very competitive and antagonistic (more on this shortly), and occasionally, the security mechanisms include means of *military force* and/or their *deterrent*. During the "Cold War" period, the North Atlantic Treaty Organization (*NATO*) and Warsaw Pact (*WarPac*) were antagonistic, and two groups considered the other as the identified enemy/opponent. This global system is actually a typical "zero-sum" game from a nation's own profit viewpoint and does not realize a regime seeking common profit (i.e., non-regime). Note that even if the situation is "zero-sum"-like, there can be alliances when the number N of actors/players is $N \geq 3$. Particularly in a diversified alliance, various rules are established in order to accelerate and maintain the cooperation, which can be a security regime. In fact, there are several

Table 8.1.   Four types of global security systems.

Threat Location	Threat Provider	
	Identified Opponent	Unidentified Opponent
External	A. Competitive Security System (Zero-Sum Game)	C. Crisis Response System (Prisoners' Dilemma, Suasion Game)
Internal	B. Controlled Power Balance (Prisoners' Dilemma)	D. Inclusive Security System (Stag Hunt Game)

rules in the NATO alliance regarding cost sharing, etc. — such as the rule not to degrade each member country's profit. Multiple regimes were created during the Cold War, since the obvious threat continued for a long period. The alliance can be considered as a *public (or club) asset* because the alliance deters external threats, and the profit/merit of the alliance should be distributed uniformly among all members (i.e., equality and non-discrimination). Even if the alliance is diversified, it still excludes non-member countries, and their collective action (in particular, military action) is against external countries, which is a strong deterrent.

China's recent territorial invasion of the Senkaku Islands in 2010 is a similar game: it happened less than a half year after a remark by (then) Japanese Foreign Minister *Katsuya Okada*, which betrayed the US–Japan Security Alliance; that "Okinawa does not have any *nuclear weapon!*" China then decided to invade these islands based on their suspicion that the Security Alliance had loosened between the Japanese Democratic Party government and the Obama's cabinet. During the earlier Liberal Democratic Party period, China had hesitated to invade these islands openly because there might have been nuclear weapon missiles in Okinawa in the US military base; that fear had been a big *deterrent* to the Chinese government.

8.1.2 *Controlled Power Balance (Type B)*

In Security System Type B, the threat (such as military forces) exists inside a group of countries, and that country is identified. An example can be found in the Conference on Security and Cooperation in Europe (*CSCE*), in which antagonistic countries tried to control their threats toward each other. CSCE began during the Cold War as a way to promote dialogue and decrease tensions between the East and West. In 1975, 35 nations signed the *Helsinki Final Act*, a politically binding declaratory understanding of the democratic principles governing relations among nations. The act contained a provision to continue regular discussions on a broad range of concerns — from migration and military security to the environment and media relations — in what became known as the *Helsinki process* [2]. Even in a global system which includes antagonistic members, there is a possibility for creating a security regime, because both sides will suffer significant damages once an actual military conflict really happens. The regime established can focus on resolving the military conflict, which is profitable to both sides. This type of regime contributes to several security functions:

(1) Crisis management — the function to escape from the actual crisis, such as military force conflicts, via various communication tools and delay lines to the final conflict decision.
(2) Trust ripening — ripening the mutual trust through accurate information/intention transfer and communication.
(3) Military power control — balancing mutual power among the antagonistic countries by eliminating a security dilemma an expansion of military forces may raise. An example of this is the agreement between the US and Russia to maintain their current number of nuclear missiles and avoid future expansion, that entered into force during the Obama presidency.

Though there remains an overall antagonistic structure (i.e., zero-sum game), we can consider a *mixed motive game* with *mixed cooperation and opposition arrangements* in the individual sections. In order to maintain this sort of regime, minimizing betrayal becomes key, and the *Prisoners' Dilemma* is the closest game that illustrates this. When you compare the payoff between "both cooperating" and "both antagonizing" action-choice combinations, the former is better and is a motivation for the regime. However, when one player betrays the regime without notifying the other side, the betrayer will receive a large payoff, while the other players will incur a deficit — a motivation for betrayal. Thus, the agreement regime, including the maintaining function, is essential.

Once the regime is established, the game structure changes drastically. This seems to transform Type B into Type A.

(1) Global ethical pressure — the agreement should be respected, or breaking the agreement degrades trust in the state in the long term.
(2) Peace and stability from military power control — breaking up the management of military power leads to an increase in international conflict.
(3) Long-term payoff — though the instantaneous payoff from the betrayal is large, the long-term payoff for not betraying the regime will be much larger.
(4) Controlled power balance — even if one player breaks the agreement and starts a military conflict, the other members will work as a whole to restrain it. This is why the system/regime is called a *controlled power balance*.

Membership into a Type B regime is usually limited to players of directly antagonistic states. However, this type of security regime is limited to the countries inside the regime, and has no negative effects on non-member states. Rather, the regime provides various positive effects to outsiders because it contributes significantly to the stabilization of the whole world system.

8.1.3 *Crisis Response System (Type C)*

In a Type C Security System, the threat (e.g., military forces) exists outside of a group of countries, and the antagonizing country is not identified. That means that there is no constant and obvious antagonistic structure in the system. An international security alliance can still exist in Type C without identifying a certain country (different from Type A). Once one of the unidentified antagonistic countries starts its military action or an action which threats international stability, the regime's members will respond to this country by means of a deterrent and/or military action. Examples of this sort of *crisis response system* include the Japan–US Security Alliance after the Cold War, and the more recent North Atlantic Treaty Organization (*NATO*). NATO members are actively sending their military forces to Afghanistan and African countries. NATO is sometimes called an *Expeditionary Alliance*.

Non-Proliferation Regimes are Type C security systems. The regimes constitute a framework for participating governments to combat multilateral problems related to export control and the proliferation of Weapons of Mass Destruction (WMDs). The regimes are comprehensive and strive to address essential threats to security. Individually, the regimes target specific threats, including chemical and biological weapons (e.g., *The Australia Group*), nuclear weapons (e.g., *Nuclear Suppliers Group*), delivery systems (e.g., *Missile Technology Control Regime* or, *MTCR*), and conventional arms (e.g., *The Wassenaar Arrangement*). By prohibiting the proliferation of weapons from "possessing" to "non-possessing" countries, the existing military power balance is maintained. When the regime identifies a specific country/territory as a threat, it turns into a Type A system, as exemplified by the *Coordinating Committee for Multilateral Export Controls* (*CoCom*) that was set up during the Cold War. However, the present Wassenaar Arrangement and MTCR are a sort of "Gentlemen's Agreement," where the possessing countries try to suppress the proliferation of military technology to "dangerous countries," which are not clearly identified until a change in the situation warrants their identification.

A crisis response system possesses an "Exclusion Principle" to non-members, which creates a sense of alienation though it contributes to global stabilization.

The Russian attitude against NATO's expansion is an example of an outsider country responding to an increasing sense of alienation. Further, it will become difficult for non-member countries to absorb new technologies from the advanced countries, which may delay their speed of economic development.

Let us consider the game models for Type C. There are two types: (1) Prisoners' Dilemma and (2) Suasion Game.

8.1.3.1 *Prisoners' Dilemma*

A regime aimed at the non-proliferation of a military weapon is considered in this example. Refer to Table 3.3(b). All the regime members possess two actions, "non-proliferation" (cooperation with the regime) and "spread" (breaking the regime by selling the weapon in question). When all the member countries cooperate with the regime by not selling any weapons, the payoff will be better than the situation in which there is no regime (i.e., motivation to comply with the regime). However, if only one member betrays the regime and sells the weapon, this country receives a large profit. Since other members will follow the betrayal for this profit, the final situation will be the lowest payoff for all members under an antagonistic situation with each other. The regime principle created is similar to Type B. An example of such a regime is the Treaty on the Non-Proliferation of Nuclear Weapons (also known as the *Non-Proliferation Treaty* or *NPT*). The NPT is an international treaty whose objective is to prevent the spread of nuclear weapons and weapons technology, to promote cooperation in the peaceful uses of nuclear energy, and to further the goal of achieving nuclear disarmament and general and complete disarmament. The Treaty entered into force in 1970, and a total of 191 states have joined it. This Treaty includes a punishment mechanism to confront any country that might betray the cause, up to the use of military force, in addition to the ethical pressure of the message that "non-proliferation is the global moral."

8.1.3.2 *Suasion Game*

The *suasion game* exists in a system where one or more countries maintain(s) a *hegemony* over other member countries. When this hegemonic country receives a large payoff from the non-proliferation treaty, the country will keep the "cooperation" action, regardless of other countries' actions. In this scenario, some countries may start betraying the system to receive a large profit. Because the hegemonic country feels frustrated with this betrayal, it may start persuading the other "possessing" countries not to spread the weapons by sharing its payoff and/or by threatening other countries, in order to keep the regime. The United States prohibited the leaking of military-related technologies to the former USSR during the Cold War by setting up the *CoCom* to regulate western alliance countries. The US occasionally used *persuasion and threats* on countries that exported advanced technologies to the communist countries.

For a long time, Soviet engineers looked to the West to help solve their most pressing military problems, and with the help of the KGB they grew adept at bypassing even the strictest export controls. In 1987, some sophisticated propeller milling equipment that enabled the Soviet Union to make much quieter submarines, slipped past the borders of Japan and Norway, and ended up in a Leningrad shipyard. When investigators probed the records of a respected Norwegian armaments maker, the state-owned Kongsberg Vapenfabrik, and a subsidiary of Japan's giant Toshiba Corporation, they began to find a trail of high-tech diversions involving sensitive equipment shipped out right under the noses of customs authorities. The US congress introduced a number of bills to bar the import of Toshiba products, and passed an unusual amendment demanding that the State Department began to negotiate for "compensation" to the US, for damages to the Navy, which claimed to have lost a long lead in its ability to detect submarines as a result of the tech leak [3].

8.1.4 *Inclusive Security System (Type D)*

In Type D systems, the threat (such as military forces) exists within the allied group of countries, but the country is not identified prior to the incident. That means that there is no obvious antagonistic structure in the system, and its members are either supporters and/or neutralists. The aim of this regime is to keep the threat inside the system, minimize threat occurrence, and deter the threat by using punishing tools. In other words, stability/peace is maintained by the group members' cooperation. The rule is equally enforced on all stakeholders/members with no discrimination. Since both the threat and the problematic countries are inside the group, this system is called an *inclusive security system* (or sometimes, "multilateralism"). This system has no negative effect on non-members. *Collective Security Alliances* indicate that military support is applied only when a member country is attacked, whether by member or non-member states (though when the attacking state is a non-member state, the system is a Type C system), and the members do not help non-member states under threat. An example of such a system is the *Multilateral Trade Agreements* (GATT/WTO), which possess a significant payoff differentiation between members and non-members.

8.1.4.1 *Concert System*

The *concert system* includes all countries between which the security system is affected mutually, with a fundamental agreement in the system's benefit, such as the stability of the present status/relationship. Further, the members agree with the cooperative action rule for an important issue under mutual understanding; without any one-way attack or threat. The power balance between large countries is not antagonistic, but cooperative. Unlike an Alliance, this concert system is meant for keeping stability within the system and between its member states, and neither benefits nor does a disservice to non-members. Historical examples of concert systems are the Congress of Vienna in 1815 after Napoleonic Wars, and the Washington Naval Treaty in the 1920s.

The *Congress of Vienna* was a conference of ambassadors of European states chaired by Austrian statesman *Klemens Wenzel von Metternich*, and held in Vienna from September 1814 to June 1815 [4]. The objective of the Congress was to provide a long-term peace plan for Europe by settling critical issues arising from the French Revolutionary Wars and the Napoleonic Wars. The goal was not simply to restore old boundaries, but to resize the main powers so they could balance each other off and remain at peace. The leaders were conservatives with little use for republicanism or revolution. France lost all its recent conquests, while Prussia, Austria, and Russia made major territorial gains. Prussia added smaller German states in the west and 40% of the Kingdom of Saxony to its territory; Austria gained Venice and much of northern Italy; Russia gained parts of Poland. The new Kingdom of the Netherlands had been created just months before, and included formerly Austrian territory that in 1830 became Belgium. The Congress of Vienna was the first of a series of international meetings that came to be known as the *Concert of Europe*, which was an attempt to forge a peaceful balance of power in Europe. It served as a model for later organizations such as the League of Nations in 1919 and the United Nations in 1945.

The *Washington Naval Treaty*, also known as the Five-Power Treaty, was a treaty among the major nations that had won World War I, which by the terms of the treaty, agreed to prevent an arms race by limiting naval construction [5]. It was negotiated at the *Washington Naval Conference*, which was held in Washington, D.C., from November 1921 to February 1922, and signed by the governments of the United Kingdom, the United States, Japan, France, and Italy. It limited the construction of battleships, battlecruisers, and aircraft carriers by the signatories. The numbers of other categories of warships, including cruisers, destroyers, and submarines, were not limited by the treaty but those ships were limited to 10,000 ton displacements.

The concert system keeps a function to suppress the intention to use military power inside the system. However, once an actual one-sided military attack is conducted, the system will collapse because the system itself does not possess the counter-military force.

8.1.4.2 *Cooperative Security System — Stag Hunt Game (Assurance Game)*

The *cooperative security system* in a Type D system is close to the concert system; the threat exists inside an allied group of countries, but the country is not identified prior to the incident, there is no obvious antagonistic structure in the system, and members are either supporters or neutralists. The differences are found in the constituting members: the concert system includes only large power states, among which the power balance is most important, while the cooperative security system includes various (large and small) states equally, and power balance is not required. All stakeholder countries (large or small) gather and discuss political and international security issues to ripen the mutual trust for preventive diplomacy and peaceful resolutions for mutual conflicts.

The Association of Southeast Asian Nations (*ASEAN*) seems to be an example of such a system. Formed in 1967 by Indonesia, Malaysia, the Philippines, Singapore, and Thailand, ASEAN aims to promote political and economic cooperation and regional stability. It has since expanded to include Brunei, Cambodia, Laos, Myanmar (Burma), and Vietnam. The ASEAN Community is composed of three pillars: The Political-Security Community, Economic Community, and Socio-Cultural Community [6].

There exist some game patterns in the cooperative security system, such as the Prisoners' Dilemma and Stag Hunt Game (or Assurance Game), of which the latter is the most popular pattern. As you have learned from Table 3.3(d), mutual cooperation generates the highest payoffs for all the players in the *stag hunt game*. All the players will receive the best results, as long as they can guarantee mutual cooperation (which is why it is called the *Assurance Game*). The regime constructed under these circumstances requires the *mutual trust* achieved by escaping misperception and by maintaining *transparency* on each state's payoff and strategic preference. Accordingly, there is no need for enforced power to establish the mutual trust; a soft structure suffices.

8.1.4.3 *Collective Security System*

The *Collective Security System* is created by expanding the previous Concert and Cooperative Security systems, by adding the military power punishment function. The sanction rule is applied without distinction, to any member state that initiates a *one-sided military threat/invasion*. The one-sided invasive action by a country is interpreted as antagonistic behavior against all the system's members; anticipating both the regime violation and the military power punishment. It is important to understand that the establishment and actual function of the collective security system requires the cooperative agreement of the large/powerful member countries with fundamental payoffs and military actions, who can practically contribute to the sanction action. In other words, a concert system within the collective security system.

The United Nations Charter is an example — though it is not a complete system of collective security, but a balance between collective actions on the one hand and the continued operation of the state system (including the continued special roles of great powers) on the other. The role of the UN and collective security is evolving given the rise of internal state conflicts. Since the end of World War II, there have been 111 military conflicts worldwide, but only nine of these have involved two or more states going to war with one another. The remainder have either been internal civil wars or civil wars in which other nations intervened in some manner. This means that collective security may have to evolve toward providing a means to ensure stability and a fair international resolution to those internal conflicts. Whether this will involve more powerful peacekeeping forces, or a larger diplomatic role for the UN, will likely be judged on a case-by-case basis.

8.1.4.4 *Ban Regime*

A ban regime is another type of general security regime without distinction that regulates the usage, possession or manufacturing of particular weapons. It is equally applied to all states who are the regime members, and provides long-term benefits to all members. The ban regime applies persuasive social pressure on non-member states to become a member, i.e., provides a positive effect to non-member states. Normally, the targets of the Ban Regime are weapons which inflict significant damage (when used) from the humanitarian standpoint — such as chemical (toxic) weapons, bio (bacterium) weapons, and anti-personnel (land) mines. The ban regimes are only for regulating these particular weapons, and not for prohibiting general warfare. The Ban Regime is therefore not a security regime, but a *humanitarian regime*. The establishment of the regime requires strong antagonistic discussions between the payoffs of the *security strategy and the humanitarian norm*.

Chemical weapons have reportedly been used by Iraq in the 1980s against the Islamic Republic of Iran. Lead by the US, the *Chemical Weapons Convention* (CWC) was adopted by the Conference on Disarmament in Geneva in 1992. The CWC allows for the stringent verification of compliance by State Parties.

The CWC opened for signature in Paris in 1993 and entered into force in 1997. The CWC is the first disarmament agreement negotiated within a multilateral frame- work that provides for the elimination of an entire category of weapons of mass destruction (WMDs) under universally applied international control.

The *Treaty on the Non-Proliferation of Nuclear Weapons* (NPT) is an international treaty whose objective is to prevent the spread of nuclear weapons and weapons technology, to promote cooperation in the peaceful uses of nuclear energy, and to further the goal of achieving nuclear disarmament and general and complete disarmament. Note that the NPT is not a global "ban" regime. The treaty recognizes five states as nuclear-weapon states: The United States, Russia, the United Kingdom, France, and China (also the five permanent members of the United Nations Security Council). Four other states are known or believed to possess nuclear weapons: India, Pakistan, and North Korea have openly tested and declared that they possess nuclear weapons, while Israel has had a policy of opacity regarding its nuclear weapons program.

8.2 Office of Naval Research Global

8.2.1 *Foundation of ONR Global*

The US Department of Navy's Naval Research Laboratory (NRL) was founded as a central research center by the famous inventor Thomas Edison. It was planned during World War I and established on July 2, 1923. After the Office of Naval Research was founded in 1946 as a bank/holding company in the Navy research community, the command assumed the responsibility of the wartime-era's Office of Scientific Research and Development liaison office in London. It aimed to identify promising research opportunities in Europe and the Middle East. By 1977, the ONR's European and Tokyo offices had combined to form the international field office with a single international S&T strategy throughout the Department of the Navy to foster international collaboration. Over the decades, ONR Global has reached out to increase and expand its cooperative activities with offices in Singapore, Tokyo, Santiago, Prague, and most recently, Sao Paulo in 2014 [7]. The US Office of Naval Research Global (ONR Global) provides worldwide S&T-based solutions for current and future naval challenges. Leveraging on the expertise of more than 50 scientists, technologists and engineers, ONR Global maintains a physical presence on five continents. The command reaches out to the broad global technical community and the operational fleet/force commands to foster cooperation in areas of mutual interest and to bring the full range of possibilities to the Navy and Marine Corps.

8.2.2 *Strategy of ONR Global*

Global S&T investment growth in a decade, from 1996 to 2006, shows a 100% growth rate; from $0.5 trillion to $1.0 trillion. In comparison with the rapid growth in China (from 4% to 13%), and in India (from 2% to 4%) in the world share, the contributions by USA, EU and Japan have decreased from 38% to 32%, from 28% to 24%, and from 16% to 13%, respectively. It is important for the US to keep the global S&T initiative moving by collaborating with the EU and Japan in order to save at least basic research investments. This is the major motivation for promoting international programs with foreign research-ers. In building and fostering international connections, ONR Global propels the execution of long-range strategic efforts that address the future needs of the naval fleet and forces, and international partners.

8.2.2.1 *Associate Director and Science Advisor Programs*

ONR Global's Associate Directors promote collaboration with international sci-entists, while Science Advisors identify fleet/force needs and implement technol-ogy solutions by seeking out practical problems experienced by the Navy fleets. Both serve as the chief of naval research's *science ambassadors abroad*. I served as a Navy "Ambassador to Japan" in this category, to promote collaborations in S&T between researchers from ONR and Japanese academic and industrial research-ers, by keeping in mind two major roles: (1) Intelligence — collecting up-to-date information on S&T development in Japan, and (2) Collaboration with the S&T Japanese community — finding researchers sympathetic with ONR's strategic developments.

8.2.2.2 *ONR Global Science Program Tools*

ONR Global sponsors international programs that address the needs of the Navy and Marine Corps and enhance the S&T priorities of ONR and the Naval Research Enterprise.

- The *Visiting Scientists Program* supports short-term travel opportunities for foreign/international scientists to the United States to share new S&T ideas or findings with the NRE (Navy Research Enterprise) that support the advancement of basic research though collaboration.
- The *Collaborative Science Program* supports foreign or international work-shops, conferences, and seminars of naval interest by providing financial support.

- The *Naval International Cooperative Opportunities in Science and Technology Program* (NICOP) provides direct research support to international scientists to help address naval S&T challenges. NICOPs support the insertion of innovative, international S&T into core ONR and Naval Research Enterprise Programs.

Interestingly, some of Japanese universities and companies prohibit military research. Thus, the ONRG distinguishes between "military research" and "basic research," and international institutes should not conduct directly military-related research, though the ONR HQ supports both basic and military-related researches for domestic researchers. This stance can be verified by the fact that the 17 researchers at NRL have been awarded the Nobel Prize so far, and ONR has been supporting researches that have led to 70 Nobel Prize winners to date. Dr. Hideki Shirakawa in Japan is one such person. He is supported by ONR, and was awarded the Nobel Prize in Chemistry in 2000. Dr. Albert Einstein, a consultant to the US Navy Bureau of Ordinance from 1943 to 1944, was also supported by ONR. If not for his discovery of the theory of relativity (which includes special relativity and general relativity), Global Positioning System (GPS) technology would not exist today. Einstein's theory of relativity enabled scientists to determine that a GPS satellite has a higher velocity than the earth as it is subject to weaker gravitational forces when in orbit, and how its atomic clock thus needs to receive calibration. This is an example of how the applications of the results of basic research are not always predictable beforehand; because on one hand, Dr. Einstein was criticized by some Japanese activists for his research's contribution to the development of the atomic bomb, while on the other hand, his discovery of relativity theory is also widely used in GPS device and/or applications that have become ubiquitous today. There is simply no guarantee that the results of basic research are not later used for developing weapons. If you think it is a good idea to boycott basic research merely because of a possibility that it may be used for weapons in the future, there would be no research or technological advancements at all. After all, technologies like the internet and GPS, which the everyman enjoys nowadays, originated from ONR studies.

8.2.2.3 *Budget Category for Military and Basic Researches*

Budget categories for basic research and military research are divided into compartments. Figure 8.1 illustrates the so-called "journey from idea to program of record and the Fleet/Force." The research classification Navy code "6.1" is

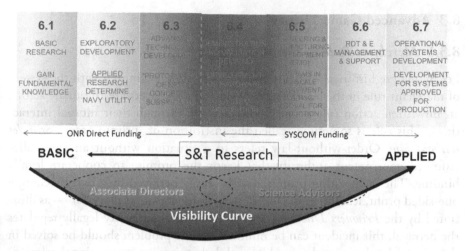

6.1	6.2	6.3	6.4	6.5	6.6	6.7
BASIC RESEARCH	EXPLORATORY DEVELOPMENT	ADVANCED TECHNOLOGY DEVELOPMENT	DEMONSTRATION AND VALIDATION (PHASE 1)	ENGINEERING & MANUFACTURING DEVELOPMENT (PHASE 2)	RDT & E MANAGEMENT & SUPPORT	OPERATIONAL SYSTEMS DEVELOPMENT
GAIN FUNDAMENTAL KNOWLEDGE	APPLIED RESEARCH DETERMINE NAVY UTILITY	"PROTO-TYPE" OF NEW CONCEPT SUBSY	DEVELOPMENT OF RECONCILING APPROACH	PLANS IN FULL SCALE EQUIPMENT INDICATING APPROVAL FOR PRODUCTION		DEVELOPMENT FOR SYSTEMS APPROVED FOR PRODUCTION

←——— ONR Direct Funding ———→ ←——— SYSCOM Funding ———→

BASIC ←—————— S&T Research ——————→ **APPLIED**

Associate Directors   Science Advisors

Visibility Curve

Fig. 8.1. Journey from idea to program of record and the fleet/force.

equivalent to "basic research." The purpose is to obtain fundamental knowledge. Engineering is "6.2" (applied research). When ONR Global supports foreign universities, we support code 6.1 and 6.2 research, but do not support military research. Research that are not confidential military research are publicized in the Broad Agency Announcement (BAA). The solicitation is publicly announced on the website, and highly competitive. Of course, we are cooperating with the Japanese Ministry of Defense in areas of research under codes "6.6" and "6.7". "6.7" refers to Operational Systems Development, including weapons, and "6.6" focuses on support (Management Support). Results for research under codes 6.6 and 6.7 are kept confidential (secret).

In the US, university researchers are encouraged to receive research funding from ONR. A large number of university professors are funded by ONR, and in the field of engineering, most of the professors receive some support from ONR. The imposition of restrictions for US universities to receive funds from ONR is rare. Of course, individual researchers are free not to receive funding from the military, or are free to only accept funding from ONR for basic research. The Applied Research Laboratory at The Pennsylvania State University is a 100% Navy-funded research institute. Johns Hopkins University is also deeply connected with the Navy, and Virginia Tech has a strong connection with the Army. Such universities' facilities conduct military-related research that lie closer to weapon applications. Otherwise, universities in general are more likely to conduct basic research.

8.3 Advanced Game Theory

8.3.1 *Merchant Rule in the Middle Ages*

Let us analyze historical institutions in this section, beginning with the system of merchant rule in the middle ages. When all the players are "rational" actors, an unwritten action rule occurs spontaneously through their mutual interactions. This rule is sometimes called the institution of "private-order" or *order without law*. Order-without-law refers to a situation without an externally-enforced government authority that keeps the promise to cooperate legally binding. The cooperation promise can be easily betrayed by a party seeking a one-sided profit, leading to the loss of payoffs for the other player(s) — as illustrated by the *Prisoners' Dilemma*. If a governmental authority legally regulates the betrayal, this incident can be minimized. The problem should be solved in the case of legal non-authority, like a global regime or international relationship. Analysis of the development of the spontaneous action rule in historical institutions is intriguing for researchers considering the methodology to generate global regimes.

In the European Middle Ages, 11th–13th centuries, merchants dealt and traded various things in the markets. Because the merchants came from a wide range of European areas, they could not bring sufficient amounts of goods for immediate sales/purchase. What they brought was just a small sample amount to one market. Then, a deal agreement was set under the promise to deliver the good(s) at some later period. When the promise was intentionally broken, a court composed of experienced senior merchants was called. The main problem in the court was the question of how they could punish the contract violators, who might not reemerge in that market. Thus, most of such courts established the following mechanism:

(1) Once a contract is violated, the report is filed to each local court.
(2) The local courts publicize the violation to other markets.
(3) The violators will be excluded from both the market in which they offended, as well as other markets.

The above situation seems to reflect the "Prisoners' Dilemma." While the violators might have received a one-time profit, they are excluded from future trading in all markets. The above mechanism functioned as "punishment" for the betrayal, and prevented successive future trading actions. Thus, the desire not to violate any contracts mutually agreed upon became the equilibrium solution in this game.

8.3.2 *Super Game Theory — Sequential Decision-Making*

The "punishment" is one solution for the "Prisoners' Dilemma." Other solutions can be found in "*Super Game* (infinite sequential games)" cases even in non-cooperative actions (i.e., situations without an explicit regulation or punishment).

8.3.2.1 *Tit-for-Tat Strategy*

In an infinitely sequential game, let us assume that the "cooperation–cooperation" action pair is taken in the first starting game. Then, after the second game, if the opponent player takes "cooperation," you will also take "cooperation," but if the opponent player takes "antagonism or betrayal," you will also take the betrayal action (i.e., *Tit-for-Tat strategy*). This tit-for-tat (TFT) strategy will reach a stable solution under a certain condition. This condition considers the total sum of future discounted profits. If the discount rate is low and the future profit weight is high in the sequential games, the "cooperation–cooperation" action pair becomes the Nash equilibrium, as we mathematically derive in the next section. Since this Nash equilibrium solution is the optimal solution, the TFT strategy is adopted as an action standard, leading to the protection of general society's (i.e., both players') profit. The TFT strategy was taken by the US government (Office of the US Trade Representative) in the 1980s in the *Super 301*; in order to promote international free trade, the US would open its market if the partner country took the open trade policy, while the US closed its market if the partner country took the closed trade policy [8].

8.3.2.2 *Principal–Agent Relationship — Trust Game*

Avner Greif's "Trust Game" theory is interesting to introduce here [4]. We suppose two players in the game; one is a principal merchant (Player I) and the other is an agent (Player II). The principal hires an agent to find out which goods to sell in a foreign state, by providing the goods to the agent first, and receiving the sales amount later. The Principal can initiate a beneficial exchange, but the Agent has the option of either fulfilling his side of the deal (cooperating) or gaining even more by reneging. If reneging occurs, the initiator of cooperation is worse off than if he had not initiated the exchange.

Let us consider this "One-Sided Prisoners' Dilemma" numerically, using a decision tree method, as illustrated in Fig. 8.2. In game form: The Principal (Player I) can either initiate exchange or not. When he does not, both Players I and II get their payoff of zero. If the Principal initiates exchange, the Agent (Player II) can decide whether or not to also exchange (i.e., cooperate by

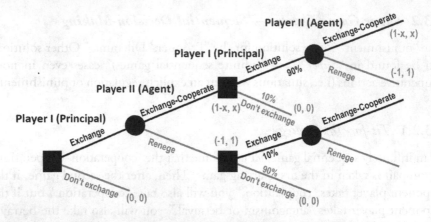

Fig. 8.2.   Decision tree for One-Sided Prisoners' Dilemma.

"fulfilling his contractual obligation"). If the exchange is efficient, it yields a total payoff of 1 that is allocated among the two players by sharing $(1 - x)$ vs. x in a way that makes both of them better off than if they do not exchange. Player I gets $(1 - x)$ (> 0) and Player II gets x (> 0). However, Player II can gain even more, specifically 1 $(> x)$, by reneging, leaving Player I with -1 (< 0), which makes him worse off than if he had not exchanged to begin with. Just from the Agent's viewpoint, the payoff for Cooperation is x, and that for Reneging is 1, which is actually better. The Principal, however, may not hire this Agent in the second round once he experiences the Agent's reneging, leading to zero earnings for the Agent in subsequent games.

Supposing that the probability of the Principal's exchange offer to the Agent in the Second Round is 90% and 10% for the Agent's Cooperation and Reneging, respectively, the expected total payoff for the Agent (Player II) for $(n + 1)$ rounds (including the first) can be estimated by:

$$\text{Cooperate–Cooperate:} \quad x + n\,0.9x$$
$$\text{Renege–Renege:} \quad 1 + n\,0.1\cdot1.$$

When $x + n0.9x > 1 + n0.1\cdot1$ (i.e., $x > \frac{1+0.1n}{1+0.9n}$), the Agent had better take the "Cooperate" strategy continuously to obtain higher profits. For a case of infinite exchanges, i.e., $n \to \infty$, even at a low share of $x > 0.11$, the Agent is better off keeping to the action to "Cooperate." As you can easily imagine, if the probability of the Principal's exchange offer for the Agent's Reneging is lower, such as 1%, the profit share will be as low as $x \approx 0.01$. In conclusion, though the Agent's one-time income for reneging is high $(=1)$, if we consider subsequent trading opportunities, the Agent is better off keeping to an honest cooperation.

8.3.2.3 *Folk Theorem*

Let us consider *super game theory* (infinitely sequential games) again, using the Prisoners' Dilemma model for a 'game' where the players are two restaurants. The two pizza restaurants, *Castle* and *Palace*, have two strategies: *high price* ($10) and *low price* ($6). The *Payoff Table* is given by Table 8.2. If the two parties can come to an agreement/collude, both will choose "High price" (Cell A), leading to a $3 million profit for both parties. Now, we consider the scenario from the *sequential decision-making* angle, by referring to Fig. 8.3: (1) Castle goes first. (2) Next, Palace makes its decision. When Castle chooses High, Palace will choose Low (Cell B), because Palace will receive $4M, which is higher than the $3M in Cell A. When Castle chooses Low, Palace will choose Low (Cell D), because $2M is higher than the $1M in Cell C. Figure 8.3(a) induces that regardless of Castle's choice, Palace's choice is always "Low," i.e., the *dominant strategy* for Palace is "Low." Consequently, in Fig. 8.3(b), when Palace inevitably

Table 8.2.   Payoff table for two restaurants with the Prisoners' Dilemma.

Castle	Palace	
	High ($10)	Low ($6)
High ($10)	A: 3, 3	B: 1, 4
Low ($6)	C: 4, 1	D: 2, 2

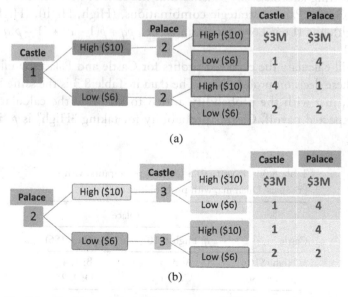

(a)

(b)

Fig. 8.3.   Decision tree for two restaurants with the Prisoners' Dilemma.

chooses Low, Castle will choose Low in the second round, because Castle will obtain only $1M if it selects "High." This solution is stable, and is called the *Nash equilibrium*, though the outcome is worse than the original High–High decision pair. Thus, we can conclude that in the *one-shot Prisoner's Dilemma*, both players cooperating is not a Nash equilibrium (if at least one of them is *rational*). The only Nash equilibrium (if both are rational) is achieved by both players defecting, which is also a mutual *mini–max* profile.

However, the situation is rather different in the *super game*. Note that the Nash equilibrium payoff in the one-shot game is lower than the Cooperation–Cooperation payoff, which is the motivation for both players to choose the "High price" option. Remember also that "TFT" is one of the strategies to maintain the "Cooperate–Cooperate" relationship. *Folk theorem* says that, in the infinitely repeated version of the game, provided players are sufficiently patient, there is a Nash equilibrium such that both players cooperate on the equilibrium path. The so-called "folk theorem" is well known, at least to experts in the game field, and considered to have established status, though work on it (in its complete form) has not been published.

In an infinitely repeated/sequential decision-making process, we consider the *decision probability* for the action items:

$$\text{Castle: Probability (High, Low)} = (p, 1 - p)$$
$$\text{Palace: Probability (High, Low)} = (q, 1 - q).$$

This setting automatically satisfies that the sum of probabilities for all the possible (Castle, Palace) strategic combinations, (High, High), (High, Low), (Low, High), and (Low, Low) equals unity, i.e., $pq + p(1 - q) + (1 - p)q + (1 - p)(1 - q) = 1$.

We will calculate the expected profits for Castle and Palace by taking into account these decision probabilities. The data in Table 8.3 is the same as that in Table 8.2, just with the probability added to facilitate the calculation. For Castle's expected payoff, Castle's probability for taking "High" is p, in which

Table 8.3.   Payoff table for two restaurants with the Prisoners' Dilemma, with probabilities.

	Palace	
Castle	q: High ($10)	$1-q$: Low ($6)
p: High ($10)	A: 3, 3	B: 1, 4
$1-p$: Low ($6)	C: 4, 1	D: 2, 2

Castle will obtain \$3M when Palace chooses "High" with the probability of q, and \$1M when Palace chooses "Low" with the probability $(1 - q)$, leading to the payoff of $p[3q + 1(1 - q)]$ (unit = \$1M). Castle's probability for taking "Low" is $(1 - p)$, in which Castle will obtain \$4M when Palace chooses "High" with the probability of q, and \$2M when Palace chooses "Low" with the probability $(1 - q)$, leading to the payoff of $(1 - p)[4q+2(1 - q)]$ (unit = \$1M). The total expected payoff for Castle is given by:

$$\text{Castle Expected Payoff} = p[3q + 1(1 - q)] + (1 - p)[4q+2(1 - q)]$$
$$= -p + 2q + 2. \tag{8.1}$$

Similarly, for Palace's payoff:

$$\text{Palace: Expected Payoff} = q[3p + 1(1 - p)] + (1 - q)[4p + 2(1 - p)]$$
$$= 2p - q + 2. \tag{8.2}$$

Equations (8.1) and (8.2), both linear relations in terms of p and q, can be visualized by changing p and q between 0 and 1, as shown in Fig. 8.4(a). Possible payoffs for Castle and Palace, by changing the probabilities p and q, are included in the parallelogram area surrounded by A(3, 3), B(1, 4), C(4, 1), and D(2, 2). Figure 8.4(b) illustrates the *folk theorem* explanation for this. Note that the Cooperate–Cooperate pair is a Nash equilibrium point, D(2, 2), where $p = q = 0$. Neither Castle nor Palace desires to lose the payoff of "2"! As long as the payoff is higher than the original Nash equilibrium amount (\$2M), both players may not complain, leading to the shaded area as the stable solution. In other words,

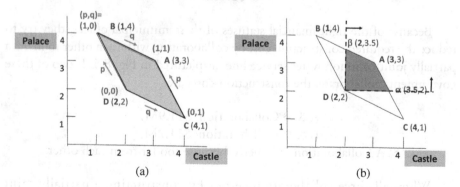

Fig. 8.4. (a) Payoffs of Castle and Palace by changing the probabilities p and q. (b) Folk theorem explanation. Note that the Cooperate–Cooperate decision pair is included in the Nash equilibrium area.

the group of Nash equilibrium solutions includes all areas in the parallelogram which originated from the D(2, 2) point (*mini–max point* or *security level*) and both players can earn higher than this level.

The solution on the lines, A − α and A − β, is called a *Pareto optimum*, the state of economic efficiency where no one player can be made better off by making the other player worse off. Note also that the *TFT strategy* guarantees the solution at the point A(3, 3) among the solutions, and the solutions at α(3.5, 2) and β(2, 3.5) require a complicated combination strategy among the "Humpty" and "Dumpty," because ($p = 0.5$, $q = 1$) and ($p = 1$, $q = 0.5$) (i.e., a tricky "High" and "Low" price 50–50 combination during an infinitely repeated game). For an infinitely repeated game, any Nash equilibrium payoff must weakly dominate the minimax payoff profile of the constituent stage game. This is because a player achieving less than his minimax payoff always has the incentive to deviate by simply playing his minimax strategy at every history.

8.3.3 *Multiple-Player Cooperative Game*

We now consider the cooperative game among multiple players, which may be a closer model to the setting in a global regime. Three municipal governments, A, B, and C are considering the construction of waterworks that draw from the same water source lake [9]. The construction costs for the individual cities are:

A: $7M,
B: $5.5M,
C: $6.5M.

Because of the tight financial statuses of these municipal offices, they try to reduce the required construction cost by collaborating with each other to make a partially joint common water service line, as pictured in Fig. 8.5. If two of these governments collaborate, the construction costs are:

A & B Collaboration: $11.9M,
B & C Collaboration: $11.2M,
C & A Collaboration: No merit, because too far from each other.

When all three collaborate together by constructing a partially-joint common water service line:

A, B & C: $17M.

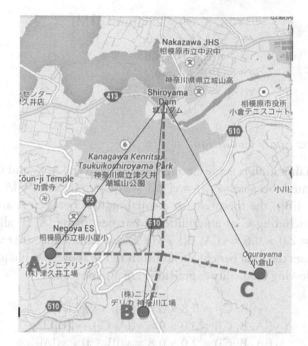

Fig. 8.5   The proposed three-way municipal government cooperation for waterworks construction.

The three governments, of course, try to minimize their required cost, as low as possible. We need to consider (1) how these three can collaborate, and (2) their share in the cost.

8.3.3.1 *Characteristic Function*

We define the group of Players as

$$N = |A, B, C|.$$

We introduce "*Characteristic Function*" as a measure of "How much can they save?" When we first consider an A and B alliance (|A, B|), we look at how constructing individually will cost $7M for A (|A|) and $5.5M for B (|B|), while the construction cost of their combined effort is $11.9M. Thus, the saving amount will be $7M + $5.5M − $11.9M = $0.6M. We denote this characteristic function as: v(|A, B|) = 0.6 [unit $1M]. Similarly, v(|B, C|) = 0.8, and v(|C, A|) = 0, because the |C, A| alliance does not reduce the construction cost. In summary, the characteristic functions for this game are:

$$v(|A, B, C|) = 2.0, [\text{unit } \$1M]$$
$$v(|A, B|) = 0.6,$$
$$v(|B, C|) = 0.8,$$
$$v(|C, A|) = 0,$$
$$v(|A|) = v(|B|) = v(|C|) = 0. \tag{8.3}$$

8.3.3.2 *Core: Share Condition*

When no collaboration is made among the three players, A, B, and C, no benefit will be generated, because $v(|A|) = v(|B|) = v(|C|) = 0$. Since $v(|A, B|) = 0.6 > 0 = v(|A|) + v(|B|)$, the incentive for a collaboration between A and B is clear. How about including C as an additional partner in the $|A, B|$ alliance? From $v(|A, B, C|) = 2.0 > 0.6 = v(|A, B|) + v(|C|)$, we can deduce that C's inclusion further increases the total benefit. In addition to the above two inequality expressions, the following inequality expressions are also obvious.

$$v(|B, C|) = 0.8 > 0 = v(|B|) + v(|C|),$$
$$v(|A, B, C|) = 2.0 > 0 = v(|A, C|) + v(|B|),$$
$$v(|A, B, C|) = 2.0 > 0.8 = v(|B, C|) + v(|A|).$$

The above inequality expressions verify that the three-player alliance is the best partnering system for minimizing the total construction cost and each individual share.

Next, we consider the share rate of the total benefit $v(|A, B, C|) = \$2.0M$ by the three players. We define a *payoff vector* as (x_A, x_B, x_C) for $|A, B, C|$. The concept of *Core* is based on the profit share principle, to which all the players are satisfactory. As an example, if $(x_A, x_B, x_C) = (1.5, 0.3, 0.2)$, do all A, B, and C players agree with this share? Because all players will at least receive some benefits in comparison with the construction cost that would have been individually borne $(v(|A|) = v(|B|) = v(|C|) = 0)$, it seems to be apparently fine. However, since the B and C two-player alliance saves \$0.8M $(v(|B, C|) = 0.8)$, the above allocation $x_B + x_C = 0.5$ is too low for players B and C, or player A receives too much x_A, leading to complaints from B and C. So, what share rate will be satisfactory for all three players? Let us consider the initial share conditions first:

$$x_A + x_B + x_C = 2.0,$$
$$x_A, x_B, x_C \geq 0. \tag{8.4}$$

Then we consider the same from the *alliance rationality*

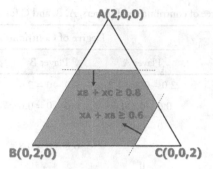

Fig. 8.6. "Core" concept of the share distribution principle.

$$x_A + x_B \geq 0.6,$$
$$x_B + x_C \geq 0.8,$$
$$x_C + x_A \geq 0,$$
$$x_A, x_B, x_C \geq 0. \tag{8.5}$$

We can visualize the payoff vector (x_A, x_B, x_C) solution that satisfies all the above conditions, as shown in Fig. 8.6. The solution is not just on one unique point, but should be in the shadowed "Core" area. Though none of the three players, A, B, or C will complain in principle, since there exist many share rate possibilities, the core approach is not convenient for practically distributing the profit in a certain unique ratio.

8.3.3.3 *Shapley Value*

Lloyd Shapley introduced the principle of share distribution in cooperative game theory, which is determined by the players' "Contribution."

As the first step, we define the *degree of contribution* of each player. The alliance/coalition |A, B| generates the payoff v(|A, B|) = 0.6. If player A drops out, the alliance |B| remains, leading to the loss of the payoff v(|A, B|) − v(|B|) = 0.6 − 0 = 0.6. Thus, we define this amount 0.6 as Player A's degree of contribution to the coalition |A, B|. Similarly, the degree of contribution of Player B can be defined by v(|A, B|) − v(|A|) = 0.6. Let us calculate the contribution of Player A in the coalition |A, B, C|, as another example. If player A drops out from the coalition |A, B, C|, the coalition |B, C| remains. Thus, the degree of the contribution can be calculated as v(|A, B, C|) − v(|B, C|) = 2.0 − 0.8 = 1.2. Table 8.4 summarizes all players' contributions for all possible coalition/alliance cases.

Second, we calculate each player's degree of contribution during the coalition/alliance processes. For example, in one scenario, Player A is the initial

Table 8.4.   Degree of contribution of players, A, B, and C for all coalition cases.

	Degree of Contribution		
	Player A	Player B	Player C
Coalition |A, B, C|	2.0 – 0.8 = 1.2	2.0 – 0 = 2.0	2.0 – 0.6 = 1.4
Coalition |A, B|	0.6 – 0 = 0.6	0.6 – 0 = 0.6	—
Coalition |A, C|	0 – 0 = 0	—	0 – 0 = 0
Coalition |B, C|	—	0.8 – 0 = 0.8	0.8 – 0 = 0.8
Coalition |A|	0	—	—
Coalition |B|	—	0	—
Coalition |C|	—	—	0

Table 8.5.   Three players' contributions for all possible coalition/alliance generation sequences.

	Contribution		
Coalition Sequence	Player A	Player B	Player C
A ← B ← C	0	0.6 – 0 = 0.6	2.0 – 0.6 = 1.4
A ← C ← B	0	2.0 – 0 = 2.0	0 – 0 = 0
B ← A ← C	0.6 – 0 = 0.6	0	2.0 – 0.6 = 1.4
B ← C ← A	2.0 – 0.8 = 1.2	0	0.8 – 0 = 0.8
C ← A ← B	0 – 0 = 0	2.0 – 0 = 2.0	0
C ← B ← A	2.0 – 0.8 = 1.2	0.8 – 0 = 0.8	0

player, then B joins, before C finally joins in to form the final team, which is denoted as A ← B ← C. For this process, the contribution of A is $v(|A|) = 0$, while that of B is $v(|A, B|) – v(|A|) = 0.6$, and that of C is $v(|A, B, C|) – v(|A, B|) = 2.0 – 0.6 = 1.4$. Thus, in the ABC process, the contributions of A, B, and C are 0, 0.6, and 1.4, respectively. As you can imagine, depending on the sequence of the three players forming the final alliance |A, B, C|, each player's degree of contribution changes.

Table 8.5 summarizes the three players' contribution for all possible coalition/alliance generation sequences.

Finally, we will introduce the *Shapley Value*. Supposing that all coalition sequences (A ← B ← C, A ← C ← B, B ← A ← C, B ← C ← A, C ← A ← B, C ← B ← A) happen under the same probability (i.e., 1/6), we calculate the expected value of the degree of each player's contribution — i.e., the "Shapley Value of each player." And, the array of Shapley values (i.e., Shapley Vector) is simply denoted as the Shapley Value. Since the Shapley values of each player can be calculated as

$$(0 + 0 + 0.6 + 1.2 + 0 + 1.2)/6 = 0.5,$$
$$(0.6 + 2.0 + 0 + 0 + 2.0 + 0.8)/6 = 0.9,$$
$$(1.4 + 0 + 1.4 + 0.8 + 0 + 0)/6 = 0.6,$$

the Shapley value is (0.5, 0.9, 0.6). Note that the Shapley value satisfies

$$x_A + x_B + x_C = 2.0,$$
$$x_A + x_B \geq 0.6, \text{ and}$$
$$x_B + x_C \geq 0.8,$$

and is included in the *Core* region. Finally, the expected share for the three players is:

$$A = 7 - 0.5 = \$6.5M;$$
$$B = 5.5 - 0.9 = \$4.6M;$$
$$C = 6.5 - 0.6 = \$5.9M.$$

8.4 International Relation Case Studies

8.4.1 *Cuban Nuclear Missile Crisis*

Essence of Decision: Explaining the Cuban Missile Crisis is an analysis of the *Cuban Nuclear Missile Crisis* authored by *Graham T. Allison*. Allison used the crisis as a case study for future studies into *governmental decision-making*. The book became the founding study of the John F. Kennedy School of Government, and in doing so, revolutionized the field of international relations [10]. We review some essentials here, cited from Alison's recent article "The Cuban Missile Crisis" in 2012 [11].

8.4.1.1 *History of the Cuban Missile Crisis*

The first Soviet Union ballistic missiles, which were Medium Range Ballistic Missiles (*MRBM*), reached Cuba on September 8, 1962, as part of the Soviet military operation called *Operation Anadyr*. The Soviet Union then built longer-range Intermediate Range Ballistic Missiles (*IRBM*) launchers in Cuba in

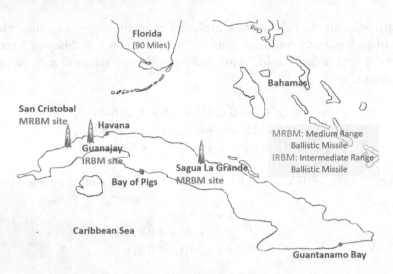

Fig. 8.7.   Map of the south of USA, and Cuba with missile bases.

October 1962. Figure 8.7 visualizes a map of the south of US, and Cuba with missile bases built. The US feared that the Soviet Union would be able to strike the US with very little warning due to the close proximity.

Then US President, *John F. Kennedy* (1917–1963), gave a speech and imposed the Naval Blockade of Cuba on October 22 1962; the following are excerpts from the speech:

First Paragraph: To halt this offensive buildup, a strict quarantine on all offensive military equipment under shipment to Cuba is being initiated.

Third Paragraph: It shall be the policy of this Nation to regard any nuclear missile launched from Cuba against any nation in the Western Hemisphere as an attack by the Soviet Union on the United States, requiring a full retaliatory response upon the Soviet Union.

Seventh and Final Paragraph: I call upon Chairman Khrushchev to halt and eliminate this clandestine, reckless, and provocative threat to world peace and to stable relations between our two nations.

8.4.1.2 *Simple Game Theory Approaches*

- Model 1: The Game of Chicken

The military executives initially suggested a military attack (air strike) against Cuba/the Soviets. This background can be explained by a game theory (*The Game of Chicken*) as follows: (1) The players are United States and Soviet Union,

Table 8.6.   Game Theory Model 1: The Game of Chicken. 4 = Best; 3 = Next best; 2 = Next worst; 1 = Worst.

| United States | Soviet Union | |
	Withdrawal	Maintenance
Blockade	3/3 Compromise	2/4 Soviet Victory, US defeat
Air Strike	4/2 US Victory, Soviet defeat	1/1 Nuclear war

Table 8.7.   Game Theory Model 2: Comments by Kennedy. 4 = Best; 3 = Next best; 2 = Next worst; 1 = Worst.

| United States | Soviet Union | |
	Withdrawal	Maintenance
Blockade	3/3 Compromise	1/4 Soviet Victory, US Capitulation
Air Strike	2/2 "Dishonorable" US action, Soviet thwarted	4/1 "Honorable" US action, Soviets thwarted"

and (2) the action items for the US include "Blockade" or "Air Strike," while those for Soviet are "Withdrawal — Removal of the nuclear missiles from Cuba" and "Maintenance — Leaving the missiles there." In order to create a payoff table for this scenario, we adopt relative numbers, which mean 4 = Best; 3 = Next best; 2 = Next worst; 1 = Worst.

The possible thinking pattern/logic by the military executives can be tabulated in the payoff table in Table 8.6. If the simple "Blockade" by the US realizes the Soviet's "Withdrawal," the resolution seems to be easier; the next best scenario is a compromise with 3/3 payoffs. However, if the "Blockade" is selected, the Soviets will probably maintain the missiles, leading to a scenario of Soviet victory and US defeat with 2/4 payoffs. Therefore, it is better for the US to take the "Air Strike" action on Cuba to force the Soviet Union to give up the setting up of nuclear missile bases. This scenario realizes US victory and Soviet defeat with 4/2 payoffs. However, there remains the worst case scenario to consider, that is in the right-bottom cell, against the US "Air Strike;" if the Soviets still keep to the "Maintenance" of their missiles, a nuclear war may ensue, which generates 1/1 payoffs.

Let us discuss the difference between the Game of Chicken and the Prisoners' Dilemma, here. When you refer to Table 3.3(b)'s "Prisoners' Dilemma," you may notice that Cells B and C provide the maximum and minimum payoff for the two players, and that though the Nash equilibrium (Cell D) provides the worse payoff for both players, it is at least still a better payoff for either player in comparison to Cells B or C. In contrast, in Table 8.6, both payoffs for the two players are the lowest in Cell D (1/1) compared to Cells B or C. Such an outcome is obvious if we consider the theory using an original chicken game drama film, *Rebel Without a Cause* (1955), by Warner Brothers, USA. In the movie, Jim and Buzz compete in the "*Chickie Run*" automobile race, in which (and Buzz explains) the rule is for "the two [drivers] to race toward the edge of the cliff, and [that] whoever jumps out of their car first is declared the 'chicken'." As the two cars speed toward the cliff, Jim tumbles out of his car, but Buzz's jacket sleeve gets caught on his door handle, preventing him from jumping out before both cars plummet to the rocky shores below. If both players had kept to the most aggressive action (i.e., keep driving) until the end, both would have experienced death, the fatal result.

- Model 2: Kennedy's Comments

President *J. F. Kennedy's* comments suggested that he did not agree with adopting the "Air Strike" option. If we create a payoff table according to his comments, Fig. 8.7 can be obtained. Kennedy's thoughts on the "Air Strike" differed significantly from Fig. 8.6's because of his *humanitarian approach* to the problem; that a US "Air Strike" on Cuba to force the Soviets to give up their nuclear missile base might only realize a "dishonorable" win for the US, with the Soviets merely thwarted (less than a defeat), and 2/2 payoffs. However, interestingly, the originally worst case scenario (in the right-bottom cell of Fig. 8.6) against the US "Air Strike," if Soviets keep their position of "Maintenance" (of missiles), that a nuclear war might commence, Kennedy thought would generate 4/1 payoffs, with the US facing an "honorable" defeat while the Soviets are dishonorably thwarted. Without seriously taking into account the action combination of "Air Strike" and "Maintenance", that is, the nuclear war, Kennedy preferred to choose "Blockade."

It is notable that, as you will see in the next section, President Kennedy initially ruled in favor of the "blockade", with "air strike" on the missile sites immediately after, if required, as an ultimatum option.

8.4.1.3 *Allison's Three Model*

The above simple game theory approaches do not give the answers to the three central questions:

(1) Why did the Soviet Union attempt to place offensive missiles in Cuba?
(2) Why did the US choose a blockade of Cuba against the Soviet missile emplacement?
(3) Why did the Soviet Union decide to withdraw the missiles?

Graham T. Allison's analysis on the above questions are introduced in the following [11].

- Why missiles in Cuba

A U-2 (the US surveillance vehicle) discovered the Soviet ballistic missile deployment in Cuba on October 14, 1962. Accordingly, on October 15–16, under shock, President Kennedy and his advisors considered the possible reasons why the Soviet Union undertook such a reckless move, and generated four possible principal hypotheses; (a) Cuban defense, (b) Cold War politics, (c) missile power, and (d) Berlin-win, trade, or trap.

The Cuban defense hypothesis: Although persuasive, explaining away Soviet nuclear missiles in Cuba with the Cuban defense hypothesis may not withstand careful examination. If *deterrence* had been the objective of this action (i.e., the prevention of a major attack), the presence of a sizeable contingent of Soviet troops would have been a better solution than the deployment of nuclear missiles. A big problem with this hypothesis is that this action actually made Cuba's position more perilous.

Cold War politics: This hypothesis considered the possibility of a Soviet objective, to demonstrate that the world balance of forces has shifted in their favor, and that the US could no longer prevent the advance of Soviet offensive power even into its own hemisphere. However, for this purpose, even a few MRBMs would have sufficed to threaten USA's entire south-eastern territory. The reason for the addition of the longer-range IRBMs and the planned deployment of submarine-launched ballistic missiles did not adequately explain/match this objective.

The missiles power hypothesis: This hypothesis suggested that the desire to tip the strategic balance of power in their favor, by increasing number of intercontinental ballistic missiles (ICBMs) and submarine-launched ballistic missiles (SLBMs), motivated the Soviet Union to this action. This might have been the major reason of this Soviet action, but there were two major objections to this hypothesis. (1) Why did Soviet premier *Nikita Sergeyevich Khrushchev* (1894–1971) feel such an extraordinary urgency to redress the strategic balance on such short notice, rather than wait to develop his ICBM force to become more formidable? And (2) why was Khrushchev willing to take such extraordinary risks to solve this problem?

The hypothesis of a *Berlin-win, trade, or trap*: A more plausible answer to the urgency of the action dawned on President Kennedy shortly afterward, i.e., that this Cuban problem was coupled with the Berlin problem. Khrushchev would use the Cuban missiles to solve the Berlin problem during his term. If the US did nothing, he would force the West out of Berlin, confident that the missiles in Cuba would deter the US from starting a war. If the Americans tried to bargain, the terms would be a trade of Cuba and Berlin. Since Berlin was more important than Cuba, that trade would also be a win for Khrushchev. If the US blockaded or attacked Cuba, Khrushchev could then use this as the excuse for an equivalent blockade or attack on Berlin. Once an attack on West Berlin by the Soviet Union happened, Kennedy thought, the US's European allies would then blame the loss of Berlin on the US, since they would not understand why the US felt the need to attack Cuba, leading to the split of the Alliance. This was, again, a win for Moscow.

• Why the American blockade

The advisors of President Kennedy differed sharply in their opinions regarding the US reaction against the Soviet/Cuban action. The *Joint Chiefs of Staff* (senior uniformed leaders in the US Department of Defense) wanted to pursue an invasion to eliminate the missile threat. They estimated that the complete invasion could be achieved in about five days from the air. In contrast, Secretary of State, *David Dean Rusk* (1909–1994), preferred pursuing diplomatic actions, at least initially; to try to get Cuban President Fidel Castro himself to push the Soviet Union out. Rusk wanted to warn Castro that the Soviets were putting Castro's regime in mortal danger and that they would happily sell him out for a victory in Berlin. This proposal, however, had no prospects; no one outside the State Department supported it. Secretary of Defense, *Robert Strange McNamara* (1916–2009), suggested that the missiles should be principally a political problem, and that the overall nuclear balance would not be significantly affected by the Soviet deployment. Therefore, McNamara was skeptical about the military action at the starting point, and argued that the option of not resorting to an "Air Strike" could be considered if the Soviet missiles in Cuba were operational and could be launched in a retaliatory strike against the US. He raised the idea of "blockading" future weapons shipments to Cuba, but his suggestion did nothing about the missiles already deployed, except to warn the Soviets not to use them.

By October 19, though the prospect of resorting to an "Air Strike" had hardened, *Robert Francis Kennedy* (1925–1968) and Rusk objected, arguing that any US surprise attack would be immoral, and likened it to the Japanese attack on Pearl Harbor. President Kennedy also mentioned that an attack on Cuba would cause a retaliatory attack on Berlin which in return would instantly present Kennedy with no meaningful military option but nuclear retaliation. Though the "blockade–negotiation" combination approach was also proposed, President Kennedy sharply ruled it out, feeling that such an action would convey to the world that the US had been frightened into abandoning its positions. He then ruled in favor of the "blockade" as an ultimatum option, urging that any air strike be limited only to the missile sites. The strike was to be ready by October 23, and Kennedy delivered his TV address on October 22. As promised in the first paragraph of his address, the Soviet ships that approached the quarantine line on October 24 were halted and turned around just before challenging it.

- Why Soviet withdraw

Because of Kennedy's "blockade" announcement, Khrushchev's fears of an immediate attack on Cuba were extinguished. Khrushchev's message to Kennedy on October 24 was defiant, and stated unyielding intention. By the morning of October 25, Khrushchev received a tough terse reply to his defiant message from Kennedy. "It was not I who issued the first challenge in this case ... I hope that your government will take the necessary action to permit a restoration of the earlier situation." With the possibility of a US attack on Cuba made obvious, Khrushchev switched to a tone of conciliation: "I was ready to dismantle the missiles to make Cuba into a zone of peace. Give us a pledge not to invade Cuba, and we will remove the missiles." He also sent instructions to accept UN Secretary-General U Thant's proposal for avoiding a confrontation at the quarantine line, promising to keep Soviet ships away from this line. Khrushchev sent another letter to Kennedy, making a more concrete offer, which acknowledged the withdrawal of Soviet missiles in Cuba, but which also demanded the withdrawal of the US missiles in Turkey.

Robert Kennedy and Theodore Sorensen composed a response to Khrushchev, which was released to the press as a public message: "The first thing that needs to be done is for work on offensive missile bases in Cuba to cease, and for all weapons systems in Cuba capable of offensive use to be rendered inoperable, under effective UN arrangements." On the Turkish question, the letter only said that "if the first proposition was accepted, the effects of such a settlement on easing world tensions would enable us to work towards a more general agreement regarding other armaments as proposed in your second letter which you made public." Because the missiles in Turkey belonged to NATO, the US could not decide on any strategy or set any deal with Khrushchev without discussing them with the members of NATO.

In Allison's summary, the blockade did not change Khrushchev's mind. It was only when it was coupled with the threat of further alternative action in the form of an airstrike that it succeeded in forcing the Soviet withdrawal of the missiles.

- Three conceptual frameworks for analyzing foreign policy

Through careful examination of the Soviet installation of missiles in Cuba, Allison introduced a complex conceptual structure. His multilayer sub-games in the player's structure are described here as the "Rational Actor" Model, "Organizational Behavior" Model, and "Government Politics" Model.

The "Rational Actor" Model: Governments are treated as the primary actor as if a *unified character*. The government examines a set of goals, evaluates them according to their utility, then picks the one that has the highest "payoff." Under this theory, Allison explains the crisis:

> *"Kennedy revealed that the Soviet Union had far fewer ICBMs than it claimed. In response, Khrushchev ordered nuclear missiles with shorter ranges installed in Cuba. Kennedy and his ExComm (Executive Committee of the National Security Council) advisors chose a "blockade" of Cuba, because it wouldn't necessarily escalate into war. Because of the mutually assured destruction by a nuclear war, the Soviets had no choice but to bow to U.S. demands and remove the weapons."*

The "Organizational Process" Model: When faced with a crisis, government leaders do not look at it as a whole, but break it down and assign it according to *pre-established organizational lines*. Because of time/resource limitations, rather than evaluating all possible actions to see which one is most suitable, leaders settle on the first adequate proposal. Under this theory, Allison explains the crisis:

> *"The Soviets mistook allowing the U.S. to easily learn of the Operation Anadyr's existence via U-2 flights over Red Army barracks. Kennedy and his advisors never considered other options besides a blockade or air strikes; while the U.S. Navy had already fallen back on the blockade as the only safe option. The Soviets simply did not have a plan to follow if the U.S. took decisive action against their missiles."*

The "Governmental Politics" Model: A nation's actions are best understood as the result of politicking and negotiations by its *top leaders*. Even if a leader holds absolute power, the leader must gain a consensus with his underlings or risk having his order misunderstood. Under this theory, Allison explains the crisis:

> *"Because of the failure of the Bay of Pigs invasion (Cuba), the Republicans made the Cuban policy into a major issue for the upcoming congressional elections. Thus, Kennedy needed to take a strong response rather than a diplomatic one. Khrushchev was losing his political power, and military leaders were unhappy with his decision to reduce the Red Army's size. Khrushchev tried to trade the issue on American missiles in Turkey. President Kennedy allowed Robert Kennedy to reach a deal with the Soviet ambassador, in which the Turkish missiles would be quietly removed several months later."*

Even though the outcome is the same, the explanation for the discussion process is rather different depending on the layer considered. *Maxwell Taylor*, Chair of the *Joint Chiefs of Staff* wanted an invasion to eliminate the missile threat, while *Dean Rusk, Secretary of State*, preferred diplomatic actions as initial issues. In contrast, *Robert McNamara*, Secretary of Defense, suggested blockading. These top leaders represent their base organizations.

8.4.2 *Operation "Tomodachi"*

International rescue missions by the US military are popular, exemplified by the Haiti Earthquake incident in January 2010. However, when Cyclone Nargis made severe landfall in Myanmar in May 2008, though 84,500 people were killed and 53,800 went missing, the Myanmar government declined US involvement in the rescue project. Though the rescue mission was primarily humanitarian in nature, Myanmar's declining of US's involvement was deeply related with the relationship between the US government and Myanmar's. In contrast, on March 11, 2011, when a 9.0 magnitude earthquake hit the eastern part of Sendai, Japan, because of the US–Japan Security Alliance originally established in 1951, collaborative work was undertaken by the US military and the Japan Self Defense Forces (JSDF). Therefore, "Operation Tomodachi" seemed to be a proper US Armed Forces operation to support the Japanese disaster relief. However, the situation was, in reality, not easy, in fact, and a kind of "game" was conducted between the then-Japanese government and the US government.

8.4.2.1 *Chronicle of Operation Tomodachi*

Over the course of the crisis, the United States and Japanese governments cooperated with one another to provide relief and aid to the affected areas of north-east Japan. *Operation Tomodachi* (Tomodachi is Japanese for "friend") was the US Armed Forces' operation that supported Japan in disaster relief following the Tohoku earthquake and tsunami. Though I knew our US Navy/Army's "friendly" operation well in the initial couple of months as a Navy Officer in ONR Global Tokyo Office, I had a sort of uncomfortable feeling about the NHK's (Japanese government-controlled broadcasting firm) news broadcasts; they had intentionally omitted/reduced broadcasting news on US soldiers' involvement in Operation Tomodachi. As a counter to this situation, I introduce the chronicling of practical "Operation" content, cited from the National Bureau of Asian Research (NBR) website. Major contributions during the initial one week are listed below for the reader's reference:

Friday, March 11

- The USS *Ronald Reagan* (nuclear power carrier, homeport at Yokosuka, Japan), stationed off the coast of Japan, works together with the Japan

Self-Defense Forces (JSDF) to provide refueling support and transport for JSDF troops.

- The USS *Tortuga*, stationed in Japan, departs for Korea to pick up *heavy-lift helicopters* to assist in relief efforts.
- US Forces Japan, based at Yokota Air Base near Tokyo, coordinates humanitarian assistance.
- US Air Force and Marine helicopter and transport aircraft are moved from Okinawa to US military bases on Honshu.

Sunday, March 13

- Two SH-60 Seahawk helicopters from the US Naval Air Facility Atsugi deliver 1,500 pounds of rice and bread to the affected region.

Monday, March 14

- The USS *Ronald Reagan* and the Carrier Strike Group move downwind from Fukushima in response to *radiological concerns* emanating from the damaged Fukushima nuclear plants. They are about 180 nautical miles away from the Fukushima nuclear complex.
- *Fifty Thousand U.S. military personnel* are operating in Japan providing disaster relief.
- P-3 Orion aircraft fly survey missions of the affected area.
- Air operations consisting of 10 helicopters from the Naval Station in Atsugi and the USS *Ronald Reagan* fly relief missions in the Minato region.

Wednesday, March 16

- A joint JSDF/US military helicopter operation to *pour water over the damaged Fukushima nuclear plants* is called off in response to the growing radiation hazard.
- Fourteen US Navy ships and their aircraft, and 17,000 sailors and Marines provide *humanitarian assistance* and disaster relief to Japan.
- The military effort includes *113 helicopter sorties and 125 fixed-wing sorties*; moving people and supplies, helping in search and rescue efforts, and delivering 129,000 gallons of water and 4,200 pounds of food.
- The USNS *Safeguard* delivers *high-pressure water pumps* to Yokota Air Base for transfer to the Japanese government for use at the Fukushima power plant.

Thursday, March 17

- Unmanned drones monitor radioactivity levels over the damaged Fukushima nuclear power plant.

- Ground Self Defense Forces mitigate the developing situation surrounding the Fukushima nuclear power plant with aid from US armed forces.
- A US Global Hawk drone and U2 spy plane measure the *radiological threat* caused by the compromised Fukushima nuclear power plant.
- Four hundred and fifty radiological and disaster-management experts prepare to offer assistance after an initial assessment by the nine-member military team from US Northern Command.
- The US Navy provides five high-capacity pumping systems to Japan's Electrical and Mechanical Engineering Group Nuclear Asset Management Department to help cool the core of the damaged No. 3 reactor at the Fukushima nuclear power plant.
- Defense Secretary Robert M. Gates authorizes up to $35 *million* in initial Defense Department funds for *humanitarian aid* to Japan.

Friday, March 18

- Four thousand additional marines and sailors with the 31st Marine Expeditionary Unit and Amphibious Squadron 11 arrive off the coast of western Japan.
- More than one dozen aircraft relocate to Naval Air Facility Misawa in northern Japan from Naval Air Facility Atsugi and Kadena Air Base in southern Japan. A total of 13 helicopters and about 500 sailors reposition to Misawa, doubling the base's population.
- Lt. Cmdr. Derek Peterson, US 7th Fleet salvage officer, visits Hachinohe, Japan, and meets with city officials to discuss the salvage capabilities the US Navy can provide in support of Operation Tomodachi.

8.4.2.2 *Diplomatic Then-Problems*

The lack of news broadcasting on US soldiers' involvement in Operation Tomodachi by NHK seemed to have originated from a sort of diplomatic friction between the US and the then-Japanese government, although a good relationship continued between the US and Japan — with the exception of this unfortunate three-year period from 2009–2012, when the Democratic Party of Japan controlled the Cabinet.

Background of Then-Cabinet: Then-prime minister (PM) *Naoto Kan*, who succeeded *Yukio Hatoyama*, was a member of the Japanese Democratic political party. Both Hatoyama and Kan had seemed to be anti-government activists in "Zengakuren" during their school days in the 1960s. The *Zengakuren* (All-Japan League of Student Self-Government) league, influenced by Marxist–Leninist political philosophy, protested strongly against the revision of the *Japan–US*

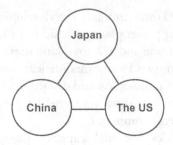

Fig. 8.8.   Political standpoint of former PM Yukio Hatoyama.

Security Treaty in 1960s. Hatoyama declared two issues during his tenure as PM (which stretched over nine months from 2009 to 2010): (1) that the US Military Base in Okinawa be moved outside the prefecture, and (2) that Japan will stand at an equal distance from China and the US, by ignoring the US–Japan Security Alliance. His standpoint is visualized in Fig. 8.8, and he and his party colleague, *Ichiro Ozawa* sent several tens of congress persons to the People's Republic of China to greet Chinese premier, *Xi Jinping*. It is important to understand this cabinet's personality from the perspective of China's territorial invasion of the Senkaku Islands in the southern Sea of Japan: in September 2010, a Chinese fishing craft intentionally collided with two Japanese coastguard patrol boats near Senkaku Island. Following the collision, coastguards arrested its crew and Captain *Zhan Qixiong*. However, to the surprise of most Japanese, the Cabinet (PM *Naoto Kan*) released all the Chinese trawler's crew members without any penalty for the territorial invasion, and, by contrast, provided a warning to the coastguard for releasing the video films, which revealed that Captain *Zhan Qixiong* had rammed his boat into the coastguard vessels, to the public. The then-Japanese Cabinet had tried to prevent this incident from going public to the Japanese people.

Problems during the Operation Tomodachi: The following dilemmas are some of Uchino's personal experiences during the Operation:

• Immediately after the Big Earthquake, the ONR Global — Asia Office at Tokyo began preparations to provide technological support to the Japanese government. We proposed to offer some robots such as (1) Unmanned Underwater Vehicles for surveying missing persons/bodies under the sea, and (2) Military iRobot (originally for ground mine search) for monitoring damages in the Nuclear Power Plant Reactors. Though we initiated communication on March 12, the reply was delayed more than seven days with a ridiculous question, "If any problem should happen to the robots, who will take responsibility?" We had not considered asking for any compensation, even if the robots totally crashed in the midst of crisis rescue. If the Japanese government had accepted our offer immediately, perhaps a couple of thousands of the bodies could have been discovered and collected, or the extent of damage by the Fukushima Daiichi Nuclear Power Plant explosion might have been diminished.

- A US Department of Defense nuclear accident expert called the Japanese government on March 12 and strongly suggested that they immediately spray water on the Daiichi Power Plant nuclear reactors. The call was broadcasted by NHK just once in the 7:00 am news on March 13. Though regular summaries of the news regarding the nuclear plants were cyclically broadcasted several times on the same day, the broadcast on the phone call never resurfaced — probably stopped by the government — and PM Kan postponed the water-cooling process longer than 72 h, leading to the fatal explosions at the power plant.

- In collaboration with the US Airforce, the Navy collected and accumulated readings on the radioactive waste spread in the air and sea, which were continuously provided to the Japanese government after March 12. However, they had not revealed these important data to the local municipal governments or citizens — action that, I believe, created much confusion in the citizen evacuation process, such as the occasional redirection of evacuees to areas of higher contamination. On the other hand, the Japanese government approached the French government for advice on the nuclear power plant accident, even though the nuclear reactors had been developed in collaboration with GE, the US, and Japan's Toshiba and Hitachi.

8.4.2.3 *Game Theory Approach*

Operation Tomodachi — The Cabinet Model: In the three models, "Rational Actor," "Organizational Behavior," and "Government Politics," the game players of Operation Tomodachi are the "US government" (a rather uniformly rational actor) and the "Then-Cabinet" (rather restricted top-leaders of Japan). The action items are rescue aid offer/request/accept and no offer or rejection of rescue aid. The Then-Cabinet's preference is, of course, no aid from the US, among which they feel a sense "victory" (score 4 as payoff) from being able to decline the US offer of Aid. The second choice (payoff 3) will be no aid from the US because of no offer from the US. The third (payoff 2) will be Then-Cabinet's reluctant acceptance of the US's aid offer, with the worst choice for the Then-Japanese Cabinet (payoff 1) being to request for rescue aid in spite of no offer from the US.

On the other hand, the US's priority was to provide rescue aid to Japan owing to the long-standing US–Japan Security Alliance. The best scenario (payoff 4) for the US is the Japanese Cabinet's request for Aid despite no offer made by the US. The second (payoff 3) being the case that an Aid offer is made by the US, and the then-Japanese Cabinet accepts it. The third (payoff 2) will be no offer from the US and, accordingly, no Aid action by the US. The worst case for the US (payoff 1) is the decline of US's Aid offer by the then- Japanese Cabinet.

The payoff table for the *Cabinet Model* is shown in Table 8.8(1). As easily seen, the *Dominant Strategy* for the US is "No Offer," while that for the then-Japanese Cabinet is "No Aid." If both parties had subscribed to this model, no Operation "Tomodachi" would have happened.

Operation Tomodachi — The General Citizen Model: A different payoff table can be created for the general Japanese citizen, because the majority of Japanese citizens (particularly people in the damaged Tohoku area) and a governmental organization, the Japan Self Defense Forces (JSDF) were welcoming of the US military's involvement in aid and rescue efforts. Because of the long-standing US–Japan Security Alliance (despite the Cabinet's preference), JSDF automatically started the collaborative rescue effort ("organizational behavior"). The average Japanese citizen generally expects a US Aid offer, which is accepted by Japan (payoff 4), followed by the case that the rescue program starts under Japanese request, despite no initial offer from the US (payoff 3). The third

Table 8.8.   Game Theory Models for "Operation Tomodachi." 4 = Best; 3 = Next best; 2 = Next worst; 1 = Worst.

United States	Japanese Cabinet	
	AID Request/Accept	No AID
(1) *The Cabinet Model*		
AID offer	3/2 Compromise	1/4 Cabinet victory, US defeat
No offer	4/1 US victory, Cabinet defeat	2/3

United States	Japanese Citizen	
	AID Request/Accept	No AID
(2) *The General Citizen Model*		
AID offer	3/4 Citizen victory	1/1 Alliance failure
No offer	4/3 US victory	2/2

choice will be "No Offer" and "No Aid." A citizen's worst case scenario is the declining of the US's Aid offer by Japanese society.

Since the US's payoff values will be the same as Table 8.8(1), a new payoff table for the General Citizen's Model can be created in Table 8.8(2). Now the *Dominant Strategy* for the US is "No Offer," while that for the general Japanese citizen is "Aid Request/Accept." So, the Nash equilibrium strategy is the scenario where Operation Tomodachi (involving the US) would start according to the Japanese Citizen's request.

Actual Operation Tomodachi: When we consider the actual Operation Tomodachi, the US offered this operation to the Japanese government, mainly due to the US's humanitarian viewpoint. The action combination of "Aid Offer from the US" and "Aid Accepted by Japan" is not the best for the US, but is the best for the Japanese Citizen and a compromised (reluctant?) action by the then-Cabinet.

8.4.2.4 *Democratic Party of Japan Afterwards*

Yukio Hatoyama succeeded *Ichiro Ozawa* as President of the Democratic Party of Japan (DPJ) before the 2009 general election, during which the DPJ swept the Liberal Democratic Party (LDP) from power in a massive landslide, winning 308 seats (out of a total of 480 seats in the lower house), while the LDP won only 119 seats — the worst defeat for a sitting government in modern Japanese history. However, because of their serious misconduct in crisis management during the Big Earthquake and the Fukushima Daiichi Nuclear Power Plant accident (including the explosion) and slow rescue actions owing to their uncooperative behavior toward the US military rescue team, the Japanese people were disappointed with the DPJ cabinet. The Japanese people were also afraid of the DPJ's too close stance to China, rather than the US. It was ousted from government by the LDP in the 2012 general election. The DPJ retained 57 seats in the lower house, and 88 seats in the upper house.

8.4.3 Cyber War

8.4.3.1 *Background of Cyber War*

The Internet is a globally extended computer network system, originally developed by the US Department of Defense in order to communicate military-related information smoothly among the US military institutes and partner industries and universities. In the middle of 1990s, the US Vice President *Albert Arnold Gore* initiated the concept of an *information highway*: analogous to the road network expansion based on the major highway. The United States needed to construct the "main artery" for information communication so that each local computer communication network could be connected worldwide. For this "main artery," the US started to use the DOD's "The *World Wide Web* (WWW) or simply Web," which is a global information medium that users can read and write via computers connected to the Internet. It is very interesting to note that the Japanese activists who are against the US military bases in Japan and R&D related to military technologies, do not hesitate to use this DOD developed "www" domain. Because of DOD's initial purpose to monitor/survey users, this web system was designed to be easily "hacked." Thus, on one hand, the US government can easily collect various intelligence and information from the world, but on the other hand, this original characteristic of the internet system is currently creating various problems against *cyber terrorists and weapons*.

Cyberattacks include any type of offensive maneuver employed by individuals or whole organizations that target computer information systems, infrastructures, computer networks, and/or personal computer devices by various means of malicious acts, usually originating from an anonymous source that either *steals*, *alters*, or *destroys* a specified target by *hacking* or *cracking* into a susceptible system. These are labeled as either a cyber campaign, cyberwarfare or cyberterrorism in different contexts. Cyberattacks can range from installing spyware on a PC, to attempts to destroy the infrastructure of entire nations. It was because of the potentially devastating consequences of cyberattacks that, in response to severe cyberattacks by China, former President *George W. Bush* declared that the US could exact revenge for cyberattacks through the use of military force.

8.4.3.2 *Case Study 1: Stuxnet*

Washington Post disclosed in 2012 that a malicious computer worm called *Stuxnet* was jointly developed by the US and Israel as a cyber weapon, aimed at sabotaging Iran's nuclear program with what would seem like a long series of unfortunate accidents. Stuxnet specifically targets Programmable Logic Controllers (PLCs), which allow the automation of electromechanical processes such as those used to control machinery on factory assembly lines, and centrifuges for separating nuclear material [12]. Exploiting four zero-day flaws, Stuxnet functions by targeting machines using the Microsoft Windows operating system and networks, then seek out Siemens Step 7 software. Stuxnet reportedly compromised Iranian PLCs, collecting information on industrial systems and causing the fast-spinning centrifuges to tear themselves apart. Stuxnet's design and architecture are not domain-specific and it could be tailored as a platform for attacking modern SCADA and PLC systems (e.g., in automobile or power plants), the majority of which reside in Europe, Japan and the US. Stuxnet reportedly ruined almost one-fifth of Iran's nuclear centrifuges.

8.4.3.3 *Case Study 2: Russo-Georgian War in 2008*

The Russo-Georgian War in 2008 focused on *a new war strategy* by practically combining *regular warfare and cyber-attacks*. An international diplomatic crisis between Georgia and Russia began when Russia announced that it would no longer participate in the Commonwealth of Independent States economic sanctions imposed on Abkhazia and established direct relations with the separatist authorities in Abkhazia and South Ossetia. The crisis was linked to the push for Georgia to receive a NATO Membership Action Plan. Increasing tensions led to the outbreak of the Russo-Georgian War in 2008. Around two months prior to the Russian Army's invasion into Abkhazia and South Ossetia in August with a large number of tank troops, Russia started cyberattacks on the Georgia internet system during the period of the Beijing Summer Olympic Games. Since Georgian economic activities were already almost dead, the invasion did not take long, demonstrating the effectiveness of Russia's new war strategy. However, in a couple of weeks, Russia officially recognized both South Ossetia and Abkhazia as independent states due to outside pressure. In response to Russia's recognition of South Ossetia and Abkhazia, the Georgian government announced that the country had cut all diplomatic relations with Russia.

8.4.3.4 *Case Study 3: Japanese Economic Recession in the 1990s*

The Japanese economic recession in the 1990s seems to also be related with cyber war. Figure 8.9 summarizes the "World Competitiveness Ranking" transition for various countries after 1991, when the IMD World Competitiveness Center started publishing World Competitive Yearbooks. In this segment, I cited the following 10 countries' ranking change trend: The US, Hong Kong (China), Singapore, Germany, Taiwan, the UK, China (Mainland), South Korea, Japan, and France. The US, Hong Kong, and Singapore are constantly top three. Mainland China moved up in world ranking after 2005. The reader should focus on the significant continuous drop of Japan from No. 1 in the early 1990s — though the technology level ranking itself does not drop much even at present. The number of the Nobel Laurates increased rather rapidly in these 10 years. Why did Japan's competitiveness decrease so quickly only several years after 1992? The reader is reminded that the GDP of Japan dropped suddenly after 1993–1994, as discussed in Fig. 2.2. The analysis in the 2003 Report on the "Necessity of Protection of Intellectual Properties for Enforcing Industrial

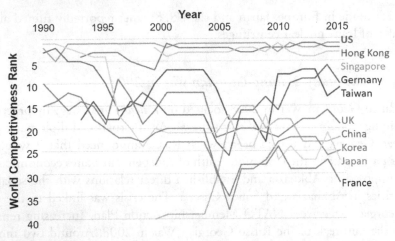

Fig. 8.9 World Competitiveness Ranking summary. Created based on World Competitiveness Yearbook from 1991 to 2015.

Competitiveness" made by the government agency, Japan Patent Office, clearly described the reason: it was caused by the *technological information leak*. Because Japan does not have a so-called *Espionage Act Regulation*, industrial spies cannot be arrested legally, nor can companies have strong regulations to ensure that employees maintain the confidentiality of industry trade secrets.

In addition to computer and internet system hacking, China and Korea adopted aggressive human recruiting strategies in the 1990s:

(1) At least a couple of hundreds of industrial researchers in major Japanese companies were recruited as weekend part-time workers in Chinese and Korean companies (private communication from my Japanese friends). The major reason many Japanese employees were receptive to the arrangement was the wage cut in their home Japanese companies due to the severe economic recession in Japan with an extraordinarily strong Japanese Yen currency.

(2) A minimum of several tens of retired Japanese professors were recruited as fixed term (3–5 years) professors in major Chinese universities. Since most of them used to work for Japanese national projects on high-tech and innovative developments, it was very easy to catch up with up-to-date Japanese technologies by hiring the retired professors in Chinese universities as special research professors, whose development ideas can be easily absorbed/stolen by young Chinese PhD students and post-doctoral fellows working with them.

Furthermore, the triggering of the Japanese Yen currency's extraordinary strength began in 1992. and was said to be manipulated by US speculators/investors via *ECHELON* to bash the strong Japanese automobile export to the US. ECHELON was a global spy system created by the US *National Security Agency* (NSA) to capture and analyze virtually every phone call, fax, email, and telex message sent anywhere in the world.

Because of the lack of cyber security strategy in Japan and almost-zero confidentiality agreements for top university researchers, key technological information were easily stolen, and Japan's global competitiveness was rapidly lost in the 1990s.

8.4.3.5 *Case Study 4: Snowden Espionage*

Edward Joseph Snowden is a computer professional, former CIA employee, and former contractor for the US Government, who presumably copied classified information from the US NSA and British Government Communications Headquarters (GCHQ) for public disclosure, such as on WikiLeaks in 2013, without prior authorization. The information revealed numerous global surveillance programs, many run by the NSA and Five Eyes, with the cooperation of telecommunication companies and European governments.

The US Department of Justice unsealed charges against Snowden of two counts of violating the *Espionage Act* and *theft* of US Government or foreign government property on June 21, 2013. Two days after, he flew to Moscow, Russia, where he reportedly remained for over a month. Later that summer, Russian authorities granted him asylum for a year, which was later extended to three years. As of 2015, he was still living in an undisclosed location in Russia while seeking asylum elsewhere.

8.5 Summary

In this chapter, we considered how to effectively establish international relationships for crisis and sustainability technologies, particularly from the S&T viewpoint. I also presented four types of international security systems: *competitive security system* (zero-sum game), *controlled power balance* (Prisoners' Dilemma game), *crisis response system* (Prisoners' Dilemma or Suasion games), and *inclusive security system* (Stag-Hunt or Prisoners' Dilemma games) — with the last system including several subsystems: *concert system, cooperative security system* (*Stag Hunt or assurance game*), *collective security system*, and *Ban regime*.

How the United States military agencies distribute their intelligence offices internationally to seek the collaborative S&T research with foreign partners, and how the R&D phases were categorized into seven types — (1) Basic research, (2) Applied research, (3) Advance technology development, (4) Demonstration and validation, (5) Engineering and manufacturing development, (6) RDT&E management and support, and (7) Operational system development — and how the first two phases can be made in academic institutes, while the last two phases are done in military enterprises, were also explained in this chapter.

Super game theory (sequential decision-making) and the *multiple-player cooperative game* were introduced as advanced game theories. I showed how Nash equilibrium solutions can be found at a certain statistical probability combinations of two action items in a super game, and how in a multiple-player cooperative game, the *Shapley value* can distribute the profit most satisfactorily to all players.

In the international relation case studies, we focused on the Cuban nuclear missile crisis, Operation "Tomodachi," and cyber warfare. I presented how, through careful examination of the Soviet installation of missiles in Cuba, Allison introduced a complex conceptual structure. His multilayer sub-games include the *Rational Actor* Model, *Organizational Behavior* Model, and *Government Politics* Model. His analysis showed how, even though the outcome (the US blockade and Russian missile base withdrawal) was the same across all the models, the explanation for the discussion process differed depending on the layer considered. I also discussed the major reasons for the delayed rescue program in the areas damaged by the tsunami areas and Fukushima Daiichi nuclear reactor's explosion, in the case study of Operation Tomodachi, and how the hesitation of the then-cabinet and Prime Minister of Japan to work with the US Army could have been due to their political background. I showed how this response may have depreciated public approval/confidence in the Democratic Party of Japan and led to its ousting from government in Japan's 2012 general election.

I also showed how cyberattacks by hacking or cracking are a new war style after the 1990s, when US Vice President *Al Gore* initiated the concept of an *information highway*; and how, because of severe cyberattacks by China, former President *George W. Bush* established that a state could take revenge for cyberattacks with military force — though we do not have a comprehensive global regime or an international agreement against cyberattacks yet. The *Stuxnet*, *Russo-Georgian War in 2008, Japanese Economic Recession in the 1990s*, and *Snowden's Espionage* were explored as case studies.

References

[1] Y. Yamamoto, Possibility of cooperative security systems — Fundamental consideration, *Int. Problems* **425**, 2–20 (1995).

[2] https://en.wikipedia.org/wiki/Helsinki_Accords.

[3] https://en.wikipedia.org/wiki/Toshiba-Kongsberg_scandal.

[4] A. Greif, The fundamental problem of exchange: a research agenda in historical institutional analysis, *Eur. Rev. Economic History* **4**(3), 251–284 (2000).

[5] https://en.wikipedia.org/wiki/Washington_Naval_Treaty.

[6] https://en.wikipedia.org/wiki/Association_of_Southeast_Asian_Nations.

[7] http://www.onr.navy.mil/science-technology/onr-global.aspx.

[8] https://en.wikipedia.org/wiki/Section_301_of_the_Trade_Act_of_1974.

[9] Hiroyuki Tezuka, *Making the Best Use of Game Theory*, Toyo-Keizai Shinpo Publ., Tokyo, Japan (2002).

[10] http://en.wikipedia.org/wiki/Essence_of_Decision.

[11] G. Allison, The Cuban Missile Crisis, Chapter 14 in *Foreign Policy — Theories, Actors, Cases*, 2nd Edition, (ed.) by S. Smith, A. Hadfield, T. Dunne, Oxford University Press, Oxford, UK (2012).

[12] https://en.wikipedia.org/wiki/Stuxnet.

CHAPTER 9

Concluding Remarks

9.1 Urgent Requirements for Crisis/Sustainability Technologies

The first quarter of 21st century seems to be the age of "Politico-Engineering," i.e., a strong political initiative is required to promote developments in engineering. The paradigm shift, from econo-engineering to politico-engineering, is much needed for the development and advancement of Science and Technology. In the 1980s, i.e., the "Bubble Economy" period, cost/performance technologies were sought, while in the last 20 years, *sustainability and crisis technologies* have been the mainstream focus. In particular, technologies in the area of piezoelectric devices have been seen in the development of (a) *non-toxic piezo-materials*, (b) *disposal technology* for existing hazardous materials using ultrasonic cavitation, (c) *reduction of contamination gas* using piezo-devices, (d) new *energy sources* (i.e., creation of piezoelectric energy harvesting), and (e) *energy-efficient piezoelectric devices*. There is an urgent need for crisis technologies in the following sensors and actuators:

(a) Accurate monitoring and surveillance techniques (e.g., vibration and pressure sensors) for earthquakes, tsunamis, typhoons, or tornados;
(b) Rescue technologies, such as autonomous unmanned underwater, aerial, land vehicles, and robots, with compact and high energy-density actuators;
(c) Detection and neutralization technologies to prevent and/or combat epidemics/infectious diseases;
(d) Nuclear power plant monitoring technologies, such as high-temperature (600°C) piezoelectric sensors, to monitor the conditions of uranium rods;
(e) Minimally destructive weapons with pin-point targets, such as laser guns and rail guns;
(f) Small energy sources for remote actuating/sensing systems and high voltage supplies with piezoelectric transformers.

9.2 Hard Power vs. Soft Power

9.2.1 *The US Monopolar State*

The military attack by the US and UK forces on Iraq beginning on March 20, 2003, provided the most impressive impact on the world power balance in the 21st century from the hard power viewpoint. The strength of USA's military force was demonstrated on daily news broadcasts of the conflict in which a large country, with territory 1.5 times the size of Japan and a population of 22 million, was occupied in less than a month in order to terminate the 24-year-long rule of a major political party lead by *Saddam Hussein*. The United States seems to be a special country; one who can attack opponent countries or a terrorist groups by itself without consulting the United Nations or their allies (sometimes even neglecting their opinions) owing to their greater military strength. However, note that the Middle-East has not been able to reach a stable or balanced status yet — more than 13 years since the first military attack.

Though the US's military power is presently the strongest in the world, it is a very young force; having developed over only 200 years since America's Independence. Its present superior status is due, in particular, to the three times world wars — World Wars I and II and the Cold War — accelerated the continuous development of war technologies, which in turn enabled American forces to avoid being defeated in those wars. In other words, the superiority of USA's military power was not accidental, but an intended product of focused economic, political, geological, and historical policies. The military power of the US was drastically enhanced during the Cold War due to their competing with the development of the Soviet Union's S&T. The key to this power enhancement in the US was to develop and adopt the most advanced military technologies (i.e., *Hard Power*). By the end of the Cold War in 1991, American hegemony (i.e., *monopoly control*) over the world system, had been established. However, on top of conventional hard power (i.e., military weapons), given the recent serious technological problem of *cyber weapons*, the US now needs to consider how to develop counter-weapons against cyberattacks to maintain the advantage.

9.2.2 *Smart Power Strategy*

The US had reached an extraordinarily high level of "Hard Power" during former President, *George W. Bush's*, tenure; and it was for this reason that the US (i.e., President Bush) attacked *Saddam Hussein's* Iraq without waiting for the consensus of the United Nations. While this "hard power" could conquer Iraqi territory in an amazingly short period of time; however, as the reader should know, this attack created serious chaos in all the territories in the Middle-East — a chaos that continues to persist 14 years on. It was on reflection of this situation that President *Barack Hussain Obama* reconsidered the US's political and security strategy in his inaugural address [1]:

> *"... Recall that earlier generations faced down fascism and communism not just with missiles and tanks, but with the sturdy alliances and enduring convictions. They understood that our power alone cannot protect us, nor does it entitle us to do as we please. Instead they knew that our power grows through its prudent use; our security emanates from the justness of our cause, the force of our example, the tempering qualities of humility and restraint. ... To the Muslim world, we seek a new way forward, based on mutual interest and mutual respect. To those leaders around the globe who seek to sow conflict, or blame their society's ills on the West, know that your people will judge you on what you can build, not what you destroy. ..."*

Joseph S. Nye, Jr., the former Assistant Secretary of Defense for International Security Affairs, proposes the *Smart Power* strategy in his book *The Future of Power* [2], where he discusses how "smart power" (achieved by coupling "hard" and "soft" power) is the key to current global diplomacy.

> *"Smart power is the combination of the hard power of coercion and payment with the soft power of persuasion and attraction. Soft power [alone] is not the solution to all problems."*

Evidence of the non-effectivity of soft power in isolation can be seen in the following examples.

Even though Kim Jong-Il, the North Korean dictator, watched Hollywood movies, it had little effect on North Korea's nuclear weapons program. And soft power had not affected much to stop Taliban's support of al-Qaeda in the 1990s. Though we understand that the smart power tactic is the best, USA's power strategy is often far too much oriented toward hard power, so much so that even US's diplomatic strategy relies on its military power. As of 2007, the annual budget for DOD (hard power) was US$500 billion, while that for the Department of State (Foreign Affairs, soft power) was US$36 billion — one order of magnitude smaller! We should consider at least a better mixture of hard and soft power. Regarding "cyberattacks," revenge using military force (as President G. W. Bush indicated) does not seem to be realistic, but developing mighty "counter-cyber weapons" to counter them will also not happen immediately. We may need to take soft power to the offending countries via persevering persuasion and attraction.

The current most serious global problem are terrorist attacks by the likes of ISIS (Islamic State of Iraq and Syria) and al-Qaeda. Against these Islamic Extremist attacks, we may not have a good strategy at this moment; no soft power may be useful, nor has any attack using purely hard power on their territories effective, because it does not reduce domestically generated terrorist actions (whether in the US, France, Japan, etc.). My generation may need to leave the development of effective anti-terrorism strategies (science and technology hardware and persuasion/attraction software) to the next generations.

9.3 Epilogue

My research target has been dramatically changed by the attacks on the World Trade Center and Pentagon on September 11, 2001. I was teaching a morning graduate class when the first warning alarm sounded in the classroom. Though no evacuation was immediately ordered by the university or the National Guard, I was wondering whether I should continue the lecture. Anyhow, during my lecture, the second warning siren rang when the Pentagon was attacked. Two days later, I started my class after offering a silent prayer for the deceased victims. After the class, two of my graduate students requested for leave of absence so as to join the Army and National Guard — promising me that they would continue the class in the next year, so that the course credit/grade could be suspended. I approved their requests. Unfortunately, they never came back to Penn State University. The younger American generation is so patriotic and ready to join the military once the necessity arises. It was because of these incidents that I decided to devote my research to crisis technologies, so as to reduce the occurrence of and/or damage by terrorist attacks.

In contrast, I was disappointed with the present younger Japanese generation's attitude toward the crisis incidents that occurred during my stay in Japan as a Navy Officer. I could not find many university professors who changed their research content, even after the incident of the Big Earthquake, towards assisting in crisis rescue; there were no developments of such technology — no robots, surveillance UAV, UUV, USV, and nuclear reactor monitoring systems — to be found. They are rather ignorant of the current state of war worldwide, i.e., "World War IV" (after the Cold War, which I count as World War III), which the US and France have officially declared against global terrorists. Because Japan may be the next target of North Korea, ISIS or China, I strongly urge Japanese engineers and researchers to prepare for this, and to work aggressively on crisis technologies in collaboration with the United States through the US–Japan Security Alliance, which the Japanese government should intentionally support.

After learning how S&T research can be adopted into crisis and sustainability technologies from this textbook, I encourage the reader to further contribute to national security in any way they can.

This book proposed a new four-Chinese-character slogan for engineering development in the era of politico-engineering: "協, 守, 減, 維 (cooperation, protection, reduction, and continuation)." Note that Prof. *Yasuhisa Hirashima* proposed a four-Chinese-character slogan in 1986 for the 21st century, i.e., "美, 遊, 潤, 創 (beautiful, amusing, tasteful, and creative) [3]"; though I still believe this will come in the future, "協, 守, 減, 維 (cooperation, protection, reduction, and continuation)" seems to be the more immediate slogan for the next 10–20 years, until we achieve global recovery in both political and economic aspects. While "beautiful, amusing, tasteful, and creative" may be effective for home appliances such as mobile devices (cell phones, tablets), it seems to still be too early for them, in general, in this period of international turmoil and terrorist attacks.

When we compare the differences between the product planning strategies for techno-, econo-, and politico-engineering, in techno-engineering (exemplified by single crystal relaxor piezoelectrics), the fundamental S&T or discovery (the 'key') comes first, followed by the development, before finally, application products are created. In econo-engineering (exemplified by piezo-multilayer components), the final product specs, in particular, in mass-production capability and manufacturing cost, come first. Based on these desired specs, manufacturing processes are developed as the key. In contrast, in politico-engineering (exemplified by Pb-free piezoelectrics), the legal regulations for the final products come first with strict constraints in terms of law-issuing date, and performance regulations (e.g., fraction of Pb content in the final product, CO_2, NO_x, or SO_x emission amounts, specifications for nuclear reactor rescue robots and new weapons). In the case of the automobile diesel injection valve, the development of a multi-layer actuator usable at an elevated temperature was the key to providing the solution; while in the RoHS regulation on the PZT, we need to discover and develop new Pb-free piezoelectric materials.

The most important issues in the age of politico-engineering are the international/global actions/motions of leading politicians on how they protect domestic industries against foreign pressure. Taking into account current domestic S&T capabilities, with an intimate collaboration with the leading domestic politicians, current global regulations need to be accepted. In other words, engineers need to instruct leading politicians on scientifically correct and up-to-date technologies! Otherwise, the domestic industrial crash may happen. As an American, I felt ashamed when I heard a comment by one of the 2016 Presidential Candidates on the "global warming" issue; that the regulation of CO_2 emissions was nonsense.

Another global discussion item can be power generation with nuclear reactors. Depending on the country's alternative energy generation capability, electric blackouts and industrial bankruptcy will happen if nuclear power plants are dismissed or completely shut down so soon just because of a general fear of nuclear power. The key is to develop the safety, security and crisis systems necessary for nuclear power plants before deciding to remove them without an alternative energy source to fill in the gap.

I sincerely hope that this book will stimulate younger researchers to consider politico-engineering, and to prepare for the paradigm shift in product planning strategy. The reader's critical comments on the book's contents are highly appreciated.

References

[1] https://www.whitehouse.gov/the-press-office/president-barack-obamas-inaugural-address.

[2] J. S. Nye, Jr., *The Future of Power*, Public Affairs, NY (2011).

[3] Y. Hirashima, *Marketing Strategy in Feeling Consumer Age*, Jitsumu-Kyoiku Publ. Co., Tokyo (1986).

Index

A

acoustic detection systems, 120
active/passive damping, 109
actors, 60
Analysis, Dissemination, Visualization,
    Insight, and Semantic Enhancement
    (ADVISE), 212
aerosol, 93
agents, 60
airport security systems, 121
Air Quality Index (AQI), 93
Al Gore or Gore, Albert Arnold, 273, 278
alliance rationality, 256
Allison, Graham T., 259
al-Qaeda, 211, 182, 284
aluminum nitride (AlN), 118, 137
american domino theory, 86
Amerithrax, 115
amusing, 52
analytic decision-making, 209
answer minimally, 197
antibodies/phages, 115
anti-cyber-attack weapons, 122
anti-earthquake technologies, 108
Anti-Personnel Mine Ban Convention
    1999, 73
anti-terrorism technologies, 120
Arab Spring, 181
Arab Winter, 181
arresters, 113
artificial fertilization, 161, 172, 175

A (second column)

ASEAN Regional Forum (ARF), 212
Asia-Pacific Economic Cooperation
    (APEC), 57, 212
Association of Southeast Asian Nations
    (ASEAN), 242
asymmetric warfare, 213
Australia group (AG), 238
average life expectancy, 95
Avigan, 116
axis of evil, 213
Azzam, Abdullah, 211

B

(Bi, Na)TiO$_3$, (BNT) piezoelectric
    ceramics, 164
bacteria, 114
ban regime, 243
batteries, 157
beautiful, 52
big science, 99
bin Laden, Osama, 182, 211
bioethanol, 145
biofuel, 91, 145, 152, 175
biometric data, 120
BLISS principle, 196
bolivia, 90
brand name value, 204
Bretton Woods system, 51
Broad Agency Announcement (BAA), 247
Bush Doctrine, 213
Bush, George W., 213, 278, 282

C

calculated risks, 210
Carter, Jimmy, 220
cavitation, 167
characteristic function, 255
Chemical Weapons Convention (CWC), 243
Chernobyl (Russia), 216
CITES, 76
civil disorder, 119
civil wars, 79, 101
CO_2 emissions, 97
coal, 142
Cold War, 44, 235–236, 282
Cold War politics, 263
Co–Li-ion types, 158
collaboration, 68
collective security alliances, 240
collective security system, 242
Commander's Critical Information Requirements (CCIRs), 209
common rail-type diesel injection valve, 168
communist Red Guards, 184
Competitive Security System, 235
competitive security system (zero-sum game), 277
Concert of Europe, 241
concert system, 241
Concrete Filled Steel Tube (CFT), 109
Conference on Security and Cooperation in Europe (CSCE), 236
Congress of Vienna, 241
containment booms, 82
continuation, viii, 52–54
contribution, 257
Controlled Power Balance, 236–237, 277
cooperation, viii, 52–54
cooperative security system, 242
Coordinating Committee for Multilateral Export Controls (CoCom), 238, 240
coordination, 63, 68
core, 257, 259
corporation, 116
coup d'état, 91
cracking, 274
creative, 52
crises, viii, 103
crisis management, 237
Crisis Response System, 238, 277
Crisis technology, 79, 179
critical infrastructure protection (CIP), 214
Cuban missile crisis, 259, 263
Cu internal electrodes, 161
cyber terrorists and weapons, 274
cyber threats, 84, 122
cyber war, 273
cyber weapon, 122, 282
cyclones, 112

D

Dalai Lama, 180, 184
Darrieus turbines, 156
Decide → Conduct → Observe → Validate, 207, 232
decision probability, 252
decision tree, 200
deep ocean currents, 155
deepwater horizon oil spill, 82
deformable mirrors, 98
degree of contribution, 257
Department of Homeland Security (DHS), 212
deterrence, 263
deterrent, 235–236
diesel additive, 145
dioxin, 167
director masao yoshida, 220
disaster, 103

disaster resilience, 215
discharging process of the reactors, 227
domestic politics, 44
dominant strategy, 62, 64, 251
drones, 221

E
earthquake, 104
Ebola, 81, 116, 183
ECHELON, 277
Edison, Thomas Alva, 132
Einstein, Albert, 133
Eisenhower, Dwight D., 86
electricity grid, 159
electromagnetic field monitoring, 104
Emergency Services Sector (ESS), 214
energy efficiency, 95
Enhanced Fujita scale (EF-Scale), 112
enormous accidents, 79, 81, 101
environmentally-friendly, 123, 137
environmental pollution, 93, 97
Environmental Protection Agency (EPA),
    82, 93, 163
epidemic/infectious disease, 79–81, 101,
    114
Espionage Act regulation, 276
European Commission, 58
European Council, 58
European Parliament, 58
European Union (EU), 58
Exclusion Principle, 238
exhaust gas regulation, 97
expeditionary alliance, 238

F
famine, 91, 161
fast neutron reactors, 149
Federal Emergency Management Agency
    (FEMA), 214
federation/confederation, 58

field manual, FM 3-0, 205
five-power treaty, 241
fixed cost, 50
flood, 112
fog of uncertainty, 208
folk theorem, 252–253
food and biofuel, 101
foreign policy analysis, 57
forum shopping, 76
fossil fuel, 152, 142
four-Rs, 215, 232
Franklin, Benjamin, 113
French Revolutionary Wars, 241
Freund, Friedemann, 105
fuel cells, 157
Fujifilm Holdings , 116
Fujita scale (F-Scale), 111
Fukushima Daiichi (Japan), 216
fungi, 114

G
game of chicken, 260
games, 60
Game Theory, 57, 76
gamma radiation, 226
gamma-ray machines, 121
gas mileage, 97
gasoline additive, 145
GDP per capita (gross domestic product
    per person), 41
general, 205
General Agreement on Tariffs and Trade
    (GATT), 60, 66
geothermal energy, 153, 175
germs, 114
gimbal, 128
global ethical pressure, 237
Global Positioning System (GPS), 246
global regime, 57
global warming, 88

governance organization, 57
government, 58
governmental decision-making, 259
governmental politics model, 265–266, 278
Go West program, 184
greenhouse effect, 88, 187
greenhouse gases (GHGs), 73, 94
"green" weapons, 123, 137
Gross Domestic Product (GDP), 42

H
hacking, 274
hard power, 60, 282
harmony game, 63, 64
hazardous waste, 92
heavier, thicker, longer, and larger, vii, 43, 54
heavy damping mechanism, 109
hegemony, 239
Helsinki Final Act, 236
Helsinki process, 237
Hendrie, Joseph, 218, 222
Homeland Security, 84, 185
Horizontal-Axis Wind Turbines (HAWTs), 155
humanitarian approach, 262
human rights, 180
hurricane, 112
Hussein, Saddam, 213, 282
hydraulic fracturing (fracking), 144
hydroelectric power, 147
hydrogen, 158
Hyper Velocity Projectile (HVP), 124
hypochlorous acid (HClO), 115

I
In Amenas hostage crisis, 91, 193
Inclusive Security System, 240, 277
Independence Game, 63–64
industrial pollution, 44–45

infectious diseases, 79, 114
information highway, 273
injection system, 174
inspection, 226
intentional accidents, 79, 82, 101
intercontinental ballistic missiles (ICBMs), 263
Interferometric-Synthetic Aperture Radar (InSAR), 107
Intermediate Range Ballistic Missiles (IRBM), 259
intermediate storage location, 230
international energy agency, 94
international relationships for crisis and sustainability developments, 235
international wars, 79
International Whaling Commission (IWC), 75
intifadas, 181
intuitive decision-making, 209
IR energy, 105
Islamic State of Iraq and Syria (ISIS), 182, 284

J
Japan Atomic Energy Research Institute (JAERI), 118
Japanese economic recession, 278
Japanese exclusive economic zone, 90
Japan Self Defense Forces (JSDF), 267
Japan–US security alliance, 238
Japan–US security treaty, 269
Jinping, Xi, 270
Joseph S. Nye, Jr., 283
Jus in Bello (international humanitarian law), 123

K
kafir, 211
Kan, Naoto, 219
Kennedy, John F., 260, 262

Kennedy, Robert Francis, 264
Khrushchev, Nikita Sergeyevich, 263
Kyoto Protocol, 53, 73, 98, 146

L

land displacement monitoring, 107
Laser Weapons System (LaWS), 123
League of Nations, 241
liberalism, 60
lighter, thinner, shorter, and smaller, vii, 46, 54
lightning, 113
lightning protector, 113
lightning rod, 113
limited political acts of terrorism, 119
lithium, 90
lithium-ion battery (LIB), 90
Little Boy atomic bomb, 216
long-term payoff, 237
low-frequency magnetic signals, 105
Lyndon Johnson's great society programs, 51

M

magenetoelectric composite sensors, 105
mathematical probability, 210
maximize personal utility, 62
medical micropump, 172
Medium Range Ballistic Missiles (MRBM), 259
membrane filtration, 163
mercury (methylmercury), 92
Metal-Oxide Varistor (MOV), 113
Metcalfe, Robert, 69
Minamata hydrate, 145
military and economic force, 59
Military Decision-Making Process (MDMP), 209
military force, 235
military power control, 237
minamata disease, 45, 92

minimax payoff, 254
mini–max point, 252, 254
Ministry of Defense, 185
Miracle of Kamaishi, 188
missiles power hypothesis, 263
Missile Technology Control Regime (MTCR), 238
mixed cooperation and opposition arrangements, 237
mixed motive game, 237
Mn–Li-ion batteries, 158
money laundering, 212
monopoly control, 282
Montreal Protocol, 73
Metal-Oxide Varistor (MOV), 137
Multilateral Trade Agreements (GATT/ WTO), 240
multimorph deformable mirror, 126
multiple-player cooperative game, 278
mutual trust, 242

N

(Na, K)NbO$_3$, (NKN) piezoelectric ceramics, 165
Nash, Jr., John Forbes, 62
nash equilibrium, 65, 67, 205, 252, 254
nash equilibrium solution, 62, 278
National Security Agency (NSA), 277
natural disaster, 79–80, 101, 103
natural gas, 142
Naval Air Systems Command (NAVAIR), 134
Naval Research Enterprise (NRE), 133
Naval Research Laboratory (NRL), 132, 133
Naval Sea Systems Command (NAVSEA), 134
Naval Surface Warfare Center (NSWC), 134
Naval Undersea Warfare Center (NUWC), 134

Navy's Naval Research Laboratory
(NRL), 244
needle-free, 174
neodymium, 90
new discoveries/inventions, 79
night vision, 127
Nixon, Richard, 86
nine principles of war, 208
non-political terrorism, 119
Non-Proliferation of Nuclear Weapons
(NPT), 60, 66, 239
non-proliferation regimes, 238
normal technologies, 79, 141
North Atlantic Treaty Organization
(NATO), 59, 66, 235, 238, 265
Nuclear and Industrial Safety Agency of
Japan (NISA), 218
Nuclear Regulatory Commission (NRC),
218
nuclear fission, 148, 149
nuclear fusion, 150
nuclear half-life, 230
nuclear power plants, 88, 152
nuclear suppliers group, 238

O

Obama, Barack Hussain, 283
ocean current, 147
OEM, 51
Office of Naval Research (ONR), 132
Office of the US Trade Representative,
249
off-shore production, 51
oil, 142
oil and civil war, 91
oil reserves, 87
Okada, Katsuya, 236
one-shot prisoner's dilemma, 252
one-sided military threat/invasion, 242
ONR global, 244

Operation Anadyr, 259
Operation Tomodachi, 267
order without law, 248
organizational behavior model, 265, 278
organizational process model, 266
Organization of the Petroleum Exporting
Countries (OPEC), 144
Ottawa Treaty, 73
oxygen, 158
Ozawa, Ichiro, 270

P

pandemic, 81
parasites, 114
pareto optimum, 254
Paris Peace Accords, 86
particulate matter (PM), 45, 93
payoff, 60
payoff table, 61, 251
Pb-free piezoelectric ceramic, 175
permanent storage, 230
persuasion and threats, 240
piezoelectric diesel injection valve, 187
players, 60
plutonium, 149
pointee-talkee, 197
political terrorism, 119
politico-engineering, viii, 54, 57,
281
pollution, 93
polychlorinated biphenyl (PCB), 167
positive-sum, 65
power, 60
power and energy, 101
power of a state/country, 59
power, profit, and norm, 76
preemptive strike strategy, 213
pre-stressed structures, 108
prisoner's dilemma, 60, 63, 66, 205, 237,
239, 248

product planning strategies, 96
profit, 60
programmable air-burst munitions (PABM), 125
Programmable Logic Controllers (PLCs), 274
Proliferation Security Initiative (PSI), 212
protection, viii, 52–54
public (or club) asset, 236
pulse-doppler radar, 111
pyroelectric effect, 128

Q

Quadratic alphabet, 197–198
quartz membrane, 115, 137
quasi-terrorism, 119
quick decision-making with insufficient information, 199
quiet killers, 162

R

radioactive water containment, 228
rail gun, 124
rapidity, 215, 232
rare-earth metals, 89, 160
rare material, 101
Rational Actor Model, 265, 266, 278
rational decision-making, 60, 61
Registration, Evaluation and Authorization of Chemicals (REACH), 163
reactor leak location detection, 229
realism, 59
real time information, 159
reduction, viii, 52–54
redundancy, 215, 232
reinforced concrete structures, 108
removal of damaged cores, 229

rendition aircraft, 214
renewable energy, 89, 95, 152
rescue robots, viii
resourcefulness, 215, 232
Restrictions on the use of certain Hazardous Substances or Restriction of Hazardous Substances Directive (RoHS), viii, 74, 93, 164
return with honor, 193
reverse analysis, 202
reverse osmosis process, 163
richter scale, 150
risk, 210
risk analysis, 202
robustness, 215, 232
Russo-Georgian war, 275
Rustum Roy, 221

S

science ambassadors abroad, 245
Sea Shepherd Conservation Society (SSCS), 75
security, 179
security cameras, 120
security level, 254
security strategy/diplomacy, 57
Senkaku Islands conflict, 86
shale gas/oil, 144
Shanghai Cooperation Organization (SCO), 212
shape memory alloys (SMAs), 137
Shapley, Lloyd, 257
Shapley value, 258–259, 279
Shia and Sunni Islamic sects, 181
Sierra Leone Civil War, 199
single crystal, 118
smart grid, 159
smart power strategy, 283
Snowden, Edward Joseph, 277
Snowden's espionage, 278

societal sustainability, ix
solar cell, 153
solar photovoltaics (solar cells), 152, 175
solar thermal energy (solar heater), 152, 175
sonochemistry, 167
Space and Naval Warfare Systems Command (SPAWAR), 134
specific power, 170
"S"-shaped growth curve, 44
Stag Hunt Game (Assurance Game), 63, 68, 242
state terrorism, 119
statistical probability, 210
status quo, 95
stored water purification and release, 228
strategic decision-making, 200
Struxnet, 274, 278
stuxnet, Russo-georgian war, 278
suasion game, 239
submarine-launched ballistic missiles (SLBMs), 263
suicide attacks, 211
super-elastic behavior of shape memory alloys (SMAs), 108
super game (infinite sequential games), 249, 252
super game theory (sequential decision-making), 251, 278
supposition, 210
surveillance camera, 128
sustainability and crisis technologies, 281
sustainability technology, 79, 88, 141
Sustainable Society, 54
Special Weapons and Tactics (SWAT), 193
swine flu, 114
System for Prediction of Environment Emergency Dose Information (SPEEDI), 223

T

tap code, 197–198
tasteful, 52
technological information leak, 276
technologies for designing/ manufacturing/ marketing, 79
territorial aggression, 79, 101
terrorist/criminal incidents, 79
tit-for-tat (TFT) strategy, 249, 254
The battle of the Sexes (Coordination Game), 67
the great wall of Japan, 110
thermal reactors, 149
Thornburgh, Richard L. "Dick", 218, 222
Three Gorges Dam, 89, 147
Three Mile Island (USA), 81, 117, 216
Tiananmen square protests, 181
Tibet autonomous region, 180
tidal wave/ocean current hydroelectricity, 152, 175
Tokyo Electric Power Company (TEPCO), 218
Operation Tomodachi, vii
tornado, 111
Tornado Alley, 111
tornadogenesis, 112
toxic materials, 92, 101
Toxic Substances Control Act (TSCA), 163
traditional security, 180
transdermal drug delivery, 174
transparency, 242
Treaty on the Non-Proliferation of Nuclear Weapons (NPT), 66, 72, 244
trichloroethylene, 167
tritium, 228
Trust Game, 249
trust ripening, 237

tungsten–bronze (TB), 165
twister, 111
typhoon, 112

U

uncertainty, 210
United Nations, 241
United Nations charter, 243
United Nations Framework Convention
   on Climate Change (UNFCCC), viii,
   53
United Nations Security Council
   (UNSC), 211
Unmanned Aerial Vehicles (UAV), 130
Unmanned Surface Vehicle (USV), 130
Unmanned Underwater Vehicles
   (UUVs), 132, 221
unmanned vehicles, 130
uranium, 149
US–Japan security alliance, 236
USS Ronald Reagan, 267
utility, 60

V

vaccine/medicine, 116
variable cost, 50
Vertical-Axis Wind Turbines (VAWTs),
   155
Viet Cong, 85
vietnamization, 86
Vietnam War, 85
virus, 114
volcanic eruption, 104
von Metternich, Klemens Wenzel, 241

W

Wallace, William S., 205
war and territorial aggression, 85

warfare gadgets, 125
warming of the earth, 74
war on terror, 211
wars, 101
warsaw pact (the warsaw treaty
   organization), 66
warsaw pact (WarPac), 235
wars/territorial invasions, 79
washington convention, 76
washington naval conference, 241
Wassenaar arrangement, 238
washington naval treaty, 241
weapons of mass destruction (WMDs),
   123, 212, 238, 244
Weiss, Joseph, 84
WikiLeaks, 277
wind turbines, 152, 155, 175
World Health Organization (WHO),
   114
world trade center, 182
world trade center attack, 82
World Trade Organization (WTO), 60,
   66
World Wide Web (WWW), 274

X

Xiaoping, Deng, 184

Y

you were doing nothing wrong, 197

Z

zero-carbon emission, 146
zero-sum, 63
zero-sum game, 64, 235
zoonotic diseases, 114

Printed in the United States
By Bookmasters

Printed in the United States
By Bookmasters